MISCELLANEOUS PUBLICATIONS
MUSEUM OF ZOOLOGY, UNIVERSITY OF MICHIGAN, NO. 119

Fishes of South Dakota

BY

REEVE M. BAILEY

AND

MARVIN O. ALLUM
South Dakota State College

ANN ARBOR
MUSEUM OF ZOOLOGY, UNIVERSITY OF MICHIGAN
JUNE 5, 1962

Paperback ISBN: 978-0-472-75012-2

CONTENTS

	PAGE
Materials and Methods	6
History of South Dakota Ichthyology	7
Collecting Stations in South Dakota	11
Annotated List of Fishes	27
Petromyzontidae	27
Acipenseridae	27
Polyodontidae	28
Lepisosteidae	29
Amiidae	30
Clupeidae	30
Salmonidae	31
Umbridae	32
Esocidae	33
Hiodontidae	33
Cyprinidae	34
Comments on the Genera *Semotilus* and *Hybopsis*	36
Catostomidae	78
Ictaluridae	87
Anguillidae	91
Cyprinodontidae	91
Gadidae	93
Gasterosteidae	93
Percopsidae	94
Serranidae	94
Centrarchidae	95
Percidae	100
Sciaenidae	107
Additional Species Reported or of Hypothetical Occurrence in South Dakota, Including Unacceptable Records	107
Hybridization	111
Some Hydrographic Interchanges that Affect Fish Distribution in South Dakota	112
Origin and Composition of the South Dakota Fish Fauna	115
Literature Cited	126

ILLUSTRATIONS

PLATE
(Plate I follows page 131)

I. Pharyngeal apparatus in *Hybognathus placitus* and *H. n. nuchalis.*

FIGURES IN TEXT

FIGURE	PAGE
1. Map of South Dakota	10
2. Distribution of *Semotilus margarita*	38
3. Distribution of *Chrosomus neogaeus*	41
4. Distribution of *Hybopsis gelida*	47
5. Distribution of *Hybopsis meeki*	49
6. Distribution of *Notropis topeka*	69
7. Regression of eye size on standard length in *Hybognathus placitus* and *H. n. nuchalis*	73

FISHES OF SOUTH DAKOTA

DESPITE a generous amount of ichthyological exploration in the northern Plains region of North America in recent decades, there is a shortage of comprehensive faunal works and synthetic analyses that assemble information into conveniently usable form. This study does not qualify as a definitive work on South Dakota fishes, but we believe that it will contribute to such an objective. We have attempted here (1) to collect and identify published records of South Dakota fishes, (2) to present keys for the identification of fishes known or expected to occur in South Dakota, (3) to record original distributional data based on 137 fish collections taken throughout the state, (4) to solve some problems involving the systematic status of certain South Dakota species, (5) to clarify or extend information on the distribution of several species in the adjoining Plains area, and (6) to determine the geographic sources of origin of South Dakota fishes and to interpret routes of postglacial dispersal. We find that the known fish fauna of the state consists of 93 species of which one is represented by two subspecies. No species is restricted to the state. Six species owe their occurrence here to introductions by man.

We wish to thank the many persons who have contributed to this study. In the field we have had the companionship and help of Marian K. Bailey, Erwin Bartel, Robert C. Gibbs, Robert E. Harris, and G. B. Spawn. We owe gratitude also to the many others who have collected South Dakota fishes that are reported here. They include Charles E. Burt, Ned E. Fogle, William A. Gosline, Carl L. and Laura C. Hubbs, George S. Myers, James T. Shields, James W. Sprague, Laurence C. Stuart, and James C. Underhill (see also list of collecting stations). Information has been supplied or specimens have been loaned by Kenneth D. Carlander, Robert E. Cleary, Frank B. Cross, Ned E. Fogle, Warren C. Freihofer, Elmer R. Liebig, George A. Moore, John B. Moyle, William R. Nelson, and James C. Underhill. Many South Dakota fish collections received by the Museum of Zoology during the period of his curatorship were identified by Dr. Carl L. Hubbs, to whom we are especially indebted. Unless indicated by subsequent systematic changes we have not re-examined such specimens. The entire manuscript was read by James T. Shields, who supplied constructive additions, particularly as concerns recent changes in the fauna resulting from impoundments and introductions. The drawings in Plate I were executed by Miss Suzanne Runyan, staff artist, Museum of Zoology.

We wish to single out for particular mention the contribution to this study of Dr. Raymond E. Johnson, now Chief, Bureau of Sport Fisheries,

U.S. Fish and Wildlife Service, who during 1939 to 1941 undertook a thorough distributional survey of Nebraska fishes (1942, University of Michigan, unpublished doctoral dissertation). Materials collected by him in waters bordering South Dakota have supplemented our own, and his findings have been utilized freely in the solution of distributional problems.

MATERIALS AND METHODS

Abbreviations for museum collections are as follows: ISU, Iowa State University; KU, University of Kansas; MCZ, Museum of Comparative Zoology, Harvard University; SDC, South Dakota State College; SDF, South Dakota Department of Game, Fish, and Parks; UMMZ, University of Michigan Museum of Zoology; and USNM, United States National Museum.

Each collecting station (p. 11) has been assigned a number, and if more than one collection was made at a station successive collections are indicated by A, B, C, etc. Stations are arranged in sequence, by drainage, beginning with the Red River drainage in the northeast corner of the state (Fig. 1). Locality and collecting data are given, and for each station the species and hybrids collected are indicated by number (as entered in the annotated list and Table 9). In the species synonymies, only original literature references to South Dakota fishes are cited; i. e., an original report by Woolman (1896) which was repeated by Cox (1897) is here attributed only to Woolman. Published records for which we have re-examined and identified specimens are indicated by asterisks. Localities of capture of each species are given by reference to the station numbers.

Measurements and fin-ray and scale counts were made according to the methods outlined by Hubbs and Lagler (1958: 19–26), unless otherwise noted. Proportional measurements were taken with a dial caliper reading to 0.1 mm., and are presented as thousandths of the standard length. Smaller parts were measured under suitable magnification. Vertebral counts were obtained from radiographs and include the hypural complex as one vertebra and the Weberian apparatus (in Cyprinidae) as four vertebrae. Specimens with visibly abnormal vertebrae are omitted.

The distributional maps (Figs. 2–6) are based largely on UMMZ specimens supplemented by published regional distribution records. All verified stations known to us are included, but we have not canvassed all possible museum holdings. Most of the keys are revised and adapted from those by Bailey (1956) and are used with the permission of the Iowa State Conservation Commission.

HISTORY OF SOUTH DAKOTA ICHTHYOLOGY

Although the French Verendrye brothers are known to have visited South Dakota as long ago as 1743, the first known records of fishes from the state are those in the journals of the Lewis and Clark Expedition, 1804 to 1806 (Burroughs, 1961). The Expedition's primary interest in fishes was as a source of food, and few accounts, including none from South Dakota, permit certain identification. Government surveys of railroad routes to the Pacific Ocean, 1853–1855, resulted in the collection of a few fishes in South Dakota taken at Ft. Pierre (presumably from the Missouri River) by Dr. John Evans, physician to the survey. They were reported by Girard (1856; 1858), and apparently constitute the earliest scientific records of South Dakota fishes. During 1876, E. D. Cope made a few fish collections from the Missouri River and "Battle Creek which empties into the Missouri" a short distance above the mouth of the Moreau River (Cope, 1879). The position of Battle Creek was given as 45° 25′ north latitude and 100° 30′ west longitude, but we cannot locate a stream there. The nearest stream of any consequence above the Moreau River is Blue Blanket Creek which enters the Missouri River from the northeast (45° 27′ 30″ N; 100° 20′ W), and we assume that this stream is Cope's Battle Creek.

Ichthyological surveys of Iowa, Minnesota, and North Dakota provide the next records of South Dakota fishes. In 1889–1891, S. E. Meek collected in the Missouri and Big Sioux rivers at Sioux City, Iowa, and the Big Sioux River at Sioux Falls, South Dakota (Meek, 1892); and in 1892 A. J. Woolman, assisted by U. O. Cox, investigated fishes of western Minnesota and eastern North Dakota (Woolman, 1896). Collection sites included several in Minnesota-South Dakota border waters as well as three in South Dakota. Later, Cox made additional collections in Minnesota and issued two reports (1896; 1897), neither of which included original records of South Dakota fishes.

In some respects, the most important work on South Dakota fishes is that by B. W. Evermann and U. O. Cox in 1892 and 1893 (Evermann and Cox, 1896). Under the direction of Prof. Evermann, the survey "for investigation and report respecting the advisability of establishing fish-hatching stations at suitable points in the states of South Dakota, Iowa, and Nebraska" and "some suitable point in Wyoming" (*op. cit.*, p. 325) made rather intensive collections in the Missouri River drainage of those states. Evermann and Cox listed 69 species of fishes from South Dakota—46 taken by them and 23 by others. Two species (not now accepted) were reported as new. In addition, Evermann described as new *Pantosteus jordani* (1893a) based on specimens taken in South Dakota. Two other papers resulted from the survey (Evermann, 1893b, and Evermann and Scovell, 1896), but these were based on the same materials.

No further collections of scientific significance were made until 1926, when the Department of Game and Fish initiated a biological survey of certain lakes of eastern South Dakota under the direction of W. H. Over, assisted by E. P. Churchill (Over and Churchill, 1927). The survey, under the direction of Dr. Churchill, was expanded in 1927 and 1928 to a statewide examination of streams and lakes, and resulted in the first publication exclusively devoted to South Dakota fishes (Churchill and Over, 1933). Eighty-one species were listed, nine of which were not taken during the survey. Two other papers were based on fishes collected by the survey (Churchill, 1927; and Hildebrand, 1932).

Aside from publications that include border waters (Eddy and Surber, 1943; Cleary, 1956; Underhill, 1957), several that treat only one or a few species (Bailey, 1959b; Bailey and Cross, 1954; Miller, 1955; Olund and Cross, 1961), and a few of restricted coverage (Allum and Hugghins, 1959; Hugghins, 1959; Moyle and Clothier, 1959; Shields, 1958a and 1958b), only one recent paper concerns the distribution of South Dakota fishes. In a survey of the Vermillion River, Underhill (1959) listed 48 species based on intensive collecting during 1955–1958.

In 1949, the Department of Game, Fish, and Parks expanded its fishery staff to include a fishery biologist, Robert C. Gibbs, who began a systematic sampling of the waters of the state, particularly the lakes. In 1950, Allum joined Gibbs and, thereafter, the collection of fishes from all waters was made an integral part of the fishery program. In the summer of 1952, Bailey joined Allum and Gibbs in making collections throughout the state, with particular attention to the Missouri River, which was then being harnessed by the first of the four dams now built or under construction. This paper is based largely on the above collections plus those in the University of Michigan Museum of Zoology and the Entomology-Zoology Department of South Dakota State College.

In the brief recorded history of South Dakota, one may note marked changes in the ichthyofauna, largely the result of intensive agriculture and introductions of fishes into new areas. Prior to the coming of the white man, the area was inhabited by a succession of Indian tribes and nations, culminating in a number of tribes of the great Sioux Nation. The Sioux, in particular, were nomadic hunters dependent on the bison and other Plains animals for their livelihood. Fishes were utilized for food when chance and circumstance allowed, as during the spring spawning migrations of suckers and other fishes, but did not constitute any major fraction of the food of the Plains dwellers. It is unlikely that these early inhabitants had any appreciable effect on the fishes of the area.

In 1861 Dakota Territory was formed to include the present states of South Dakota, North Dakota, Montana, and Wyoming. Homesteading commenced in South Dakota near Vermillion in 1863. Then the plowing of the prairies began, and with that the change in the character of the streams was rapidly accelerated, particularly in the eastern third of the state. Civilization brought other changes as well, and when Evermann and Cox (1896: 336) observed Whitewood Creek in the Black Hills in 1892, it was already ". . . ruined by the tailings from the numerous stamp mills." Despite a few such isolated cases, industrial and municipal pollution of streams has probably contributed little to faunal change in the state.

One of man's activities most responsible for changes in the distribution of fishes is that of the introduction of exotic species (p. 125) and transplantation of native species. In 1883, the Territorial Legislature provided for the appointment of a "Fish Commissioner" whose duties included the planting of fish fry received from the U.S. Commissioner of Fisheries, and the introduction of the carp, *Cyprinus carpio*, began about 1885 (Hipschman, 1959; 23, 28). Although Evermann and Cox took no carp in their survey during 1892 and 1893, they reported (1896: 347) that carp had been planted in two lakes in north-central Nebraska [Brown County] as early as 1887. The effect of this hardy, competitive fish on the native fishes is difficult to evaluate.

Transplantation has largely affected the distribution of the food and game fishes, although bait and forage species have been included. The original range of many species cannot, therefore, be surely defined. As long ago as 1892, Evermann and Cox (1896: 417) recognized this problem and noted: "Both species of *Pomoxis* [crappies] are being extensively introduced into the waters of Kansas, Nebraska, and South Dakota, and it is not easy to determine definitely the natural western limit of either."

Stream impoundments and the construction of farm and ranch ponds have modified the distribution of fishes, particularly of the sport species. Even greater modification may be predicted, chiefly in the composition of the fish fauna of the Missouri River during its transition from a great river of rather steep gradient, high turbidity, and wide fluctuations in volume to a series of clear reservoirs separated by short stretches of river in which flow is controlled. We note with gratification that fishery biologists of the South Dakota Department of Game, Fish, and Parks are tracing these changes qualitatively and quantitatively.

Nature has played her part in the changes in fish distribution, particularly during the drought period of the 1930's. During the worst of the dry

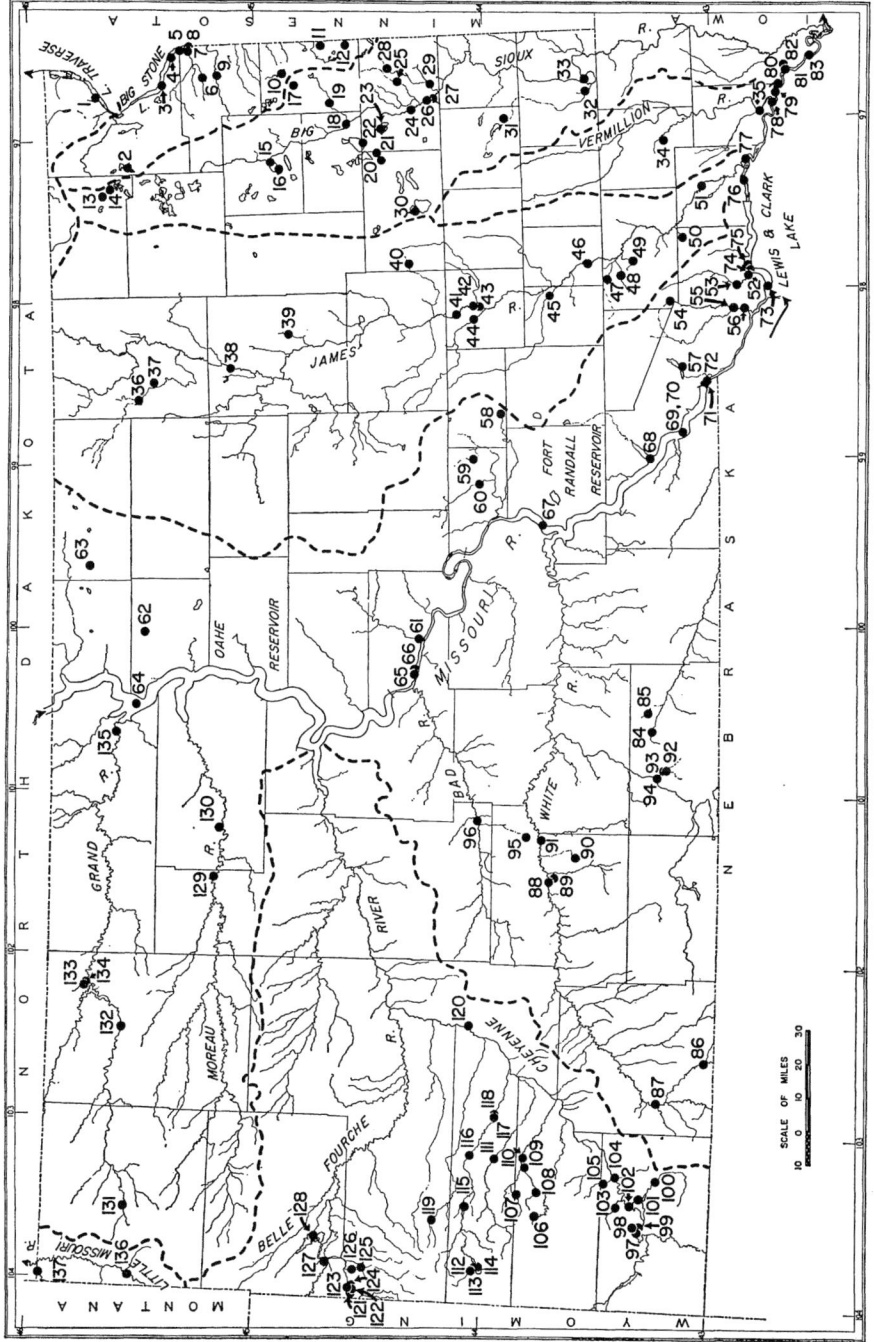

FIG. 1. South Dakota showing principal streams and drainage areas, and fish collecting stations.

years, only six lakes in South Dakota—Big Stone, Cochrane, Enemy Swim, Kampeska, Pickerel, and Punished Woman—are known to have supported game fishes, and a few more are believed to have sustained some fish life. Some of the species reported by Churchill and Over (1933), but not taken by us, may have been extirpated during this critical period.

The trend toward urbanization, with its encroachment on nature, is less evident in South Dakota than elsewhere, and the population of the state has shown little increase. It seems likely, therefore, that barring the introduction of additional exotic species and the possible passing of a few rare species no major ichthyological changes may be expected in the near future.

COLLECTING STATIONS IN SOUTH DAKOTA

Original field collection numbers are entered in parentheses after the names of the collectors.

RED RIVER OF THE NORTH BASIN

STATION 1.—Lake Traverse, T. 126 N, R. 49 W, Roberts Co., Sept. 23, 1949, R. C. Gibbs and D. Allen (G24–49). Lake Traverse is a large (11,000 acres), shallow (maximum depth 16 feet) prairie lake and is the main headwater of the Red River of the North. A detailed description of the lake was given by Moyle and Clothier (1959: 178–79). SDF. Species: 16, 44, 48, 61, 77, 80, 81, 83, 84, 87, 92.

MINNESOTA RIVER BASIN

STATION 2.—Drywood L., T. 125 N, R. 52 W, Roberts Co., Aug. 30, 1950, Marvin O. Allum, Alvin Peterson, and Erwin R. Bartel (FR2–51). Drywood Lake is one of a chain of small, shallow prairie lakes, all subject to winterkill. SDF. Species: 48, 61, 84, 87.

STATION 3.—Big Stone L., Hartford Beach, T. 122 N, R. 48 W, sec. 4, Roberts Co., Aug. 26, 1952, Gibbs and Bartel (B52–73). Big Stone Lake is a large, narrow lake (about 25 miles long and up to 1.5 miles wide), and is relatively shallow (maximum depth about 30 feet). The shoreline is largely sand, gravel, and/or boulders. Silt is common in deeper water. UMMZ 167128–50. Species: 14, 20, 25, 29, 32, 35, 37, 39, 41a, 44, 47, 48, 49, 58, 73, 75, 80, 83, 84, 87, 91, 92, 93.

STATION 4.—Big Stone L., T. 122 N, R. 47 W, sec. 25, Roberts Co., Aug. 26, 1952, Gibbs and Bartel (B52–72). UMMZ 167063–75, 167126–27. Species: 16, 20, 25, 29, 35, 47, 48, 58, 61, 77, 79, 80, 83, 87, 91, 93.

STATION 5.—Big Stone L., T. 122 N, R. 46 W, sec. 31, Roberts Co., Aug. 20, 1949, Gibbs and Allen (G23–49). SDF; UMMZ 163837. Species: 20, 29, 32, 37, 39, 44, 47, 48, 61, 80, 91.

STATION 6.—[S Br.] Whetstone Cr., inlet to Big Stone L., Milbank, Grant Co., July 6, 1956, J. C. Underhill, H. McAllister, and H. Orr (56–016). UMMZ 171145. Species: 92.

STATION 7.—Whetstone Cr., 1 mi. W Big Stone City, T. 121 N, R. 46 W, sec. 18, Grant Co., Aug. 26, 1952, Allum, R. M. Bailey, and R. E. Harris (B52–63). Water almost clear, visibility 24 inches, 78° F.; bottom gravel, rubble, sand; some algae on rocks in riffles; current slight to swift; depth to 3 feet; width 12 to 25 feet. UMMZ 166890–909. Species: 20, 25, 29, 33, 35, 39, 41a, 47, 49, 53, 56, 57, 58, 66, 82, 83, 84, 86, 89, 91; hybrid 25 $\times$ 35.

STATION 8.—Whetstone Cr., Big Stone City at U.S. 12, T. 121 N, R. 46 W, sec. 17, Grant Co., Aug. 26, 1952, Gibbs and Bartel (B52–71). Water clear, 79° F.; bottom gravel, sand, silt, clay; some narrow-leafed *Potamogeton;* depth to 7 feet. UMMZ 166992–167024. Species: 14, 16, 20, 25, 29, 32, 35, 39, 41a, 44, 47, 48, 49, 51, 53, 56, 57, 58, 61, 65, 66, 73, 75, 77, 79, 80, 82, 83, 84, 87, 88, 91, 93.

STATION 9.—N Fk. Yellowbank R., 3.5 mi. S Milbank at U.S. 77, T. 120 N, R. 47 W, sec. 30, Grant Co., Aug. 26, 1952, Allum, Bailey, and Harris (B52–62). Water murky, 73° F.; bottom gravel, sand, soft mud; vegetation slight; no current; depth to 2 feet; width to 15 feet. UMMZ 166874–89. Species: 20, 25, 29, 35, 39, 44, 47, 48, 49, 58, 61, 81, 83, 86, 87, 88, 91.

STATION 10.—Lake Alice, T. 116 N, R. 49 W, sec. 1, Deuel Co., Aug. 17, 1949, Gibbs and Allen (G6–49). SDF. Species: 48, 61, 73.

STATION 11.—Lake Cochrane, T. 114 N, R. 47 W, sec. 4, Deuel Co., Aug. 17, 1949, Gibbs and Allen (G3–49). SDF. Species: 35, 39, 48, 49, 69, 87, 92.

STATION 12.—Fish L., T. 113 N, R. 47 W, Deuel Co., Aug. 17, 1949, Gibbs and Allen (G4–49). SDF. Species: 48, 61.

MISSOURI RIVER BASIN

Big Sioux River Drainage

STATION 13.—Clear Lake, SE Lake City, T. 126 N, R. 54 W, Marshall Co., June 27, 1950, Gibbs, Merlon Dahl, and Peterson (G51–50). SDF. Species: 48, 87.

STATION 14.—Cottonwood Lake, T. 126 N, R. 53 W, Marshall Co., Sept. 26, 1949, Gibbs and Allen (G29–49). SDF. Species: 20, 48, 77, 81, 84, 87, 92.

STATION 15.—Lake Kampeska at outlet, T. 117 N, R. 53 W, sec. 15,

Codington Co., Aug. 18, 1949, Gibbs and Allen (G7–49). SDF; UMMZ 163836. Species: 18, 32, 41a, 48, 61, 65, 74, 75, 81, 83, 86, 90, 91.

STATION 16.—Lake Kampeska at Stony Point Resort, T. 117 N, R. 53 W, sec. 29, Codington Co., Aug. 18, 1952, Gibbs and Allum. Water clear; bottom sand, gravel and rubble beach; no rooted vegetation; depth to 3.5 feet. SDF. Species: 20, 32, 35, 37, 39, 41a, 48, 49, 61, 74, 75, 80, 84, 87, 90, 91.

STATION 17.—Ketcham (South Coteau) L., T. 116 N, R. 49 W, Deuel Co., Aug. 17, 1949, Gibbs and Allen (G5–49). SDF. Species: 48, 61, 87.

STATION 18.—Big Sioux R., 2.5 mi. S Dempster, T. 113 N, R. 51 W, sec. 15, Hamlin Co., Apr. 21, 1956, Allum (SC3–56). Water clear; bottom sand, gravel, some rubble in current; some attached filamentous algae; current sluggish to moderate; depth to 3 feet; width to 25 feet. SDC. Species: 20, 29, 35, 39, 48, 58, 61, 91.

STATION 19.—Hidewood Cr., T. 114 N, R. 50 W, sec. 16, Deuel Co., Apr. 21, 1956, Allum (SC2–56). Water clear, 58° F.; bottom sand, gravel, some rubble; attached filamentous algae; current sluggish to moderate; depth to 2 feet; width to 15 feet. SDC. Species: 20, 39, 48, 49, 73, 91, 92.

STATION 20.—Badger L., 2 mi. SE Badger, T. 112 N, R. 53 W, Kingsbury Co., Aug. 19, 1949, Gibbs and Allen (G10–49). SDF. Species: 48, 81, 87.

STATION 21.—Thisted L., 3 mi. ESE Badger, T. 112 N, R. 53 W, Kingsbury Co., Aug. 19, 1949, Gibbs and Allen (G9–49). SDF. Species: 16, 48, 81, 87.

STATION 22.—Lake Poinsett, south shore, T. 112 N, R. 52 W, sec. 4, Brookings Co., Aug. 16, 1949, Gibbs and Allen (G2–49). SDF; UMMZ 163835. Species: 32, 37, 48, 81, 86, 87, 91.

STATION 23.—Tetonkaha L., 5 mi. W Bruce, T. 111 N, R. 51 W, sec. 5, Brookings Co., Aug. 16, 1949, Gibbs and Allen (G1–49). SDF. Species: 42, 48, 61, 73, 81, 87, 91, 92.

STATION 24.—Big Sioux R., 2 mi. E Volga, at U.S. 14, Brookings Co., July 15, 1950, M. K. and R. M. Bailey (B50–38). Water murky, visibility about 12 inches, 72° F.; bottom mostly mud, some sand and gravel; no vegetation; current sluggish to moderate; depth to 4 feet; width 30 to 40 feet. UMMZ 161924–36. Species: 20, 37, 39, 40, 41a, 47, 48, 51, 58, 61, 81, 87, 91.

STATION 25.—Spring hole and creek trib. to Six-mile Cr., 6 mi. NE Brookings, T. 111 N, R. 49 W, sec. 32, SE ¼, Brookings Co.

A: Jan. 1947, G. B. Spawn and Allum (SC47–1). Water clear, cold; bottom silt over sand; no vegetation; current slight; depth to 3 feet; width 2 to 12 feet. SDC; UMMZ 173146. Species: 13 (others not recorded).

B: Aug. 26, 1952, Allum, Bailey, and Harris (B52–61). Water clear, 57°

F.; bottom soft silt, gravel in creek, vegetation abundant; current slight; depth to 2 feet; width 2 to 12 feet. UMMZ 166863–73. Species: 14, 20, 22, 29, 39, 44, 48, 49, 58, 73, 91, 92.

STATION 26.—Big Sioux R., 5 mi. SW Brookings, T. 109 N, R. 50 W, secs. 9, 16, Brookings Co. Water murky to turbid; bottom sand, gravel, silt; vegetation none; current sluggish to moderate; depth to 4 feet; width 30 feet.

A: May 16, 1956, Allum and students (SC5–56). SDC. Species: 32, 35, 40, 41a, 47, 48, 58, 61, 81.

B: May 24, 1956, Allum and students (SC6–56). SDC. Species: 16, 20, 32, 35, 40, 41a, 48, 49, 61.

C: Aug. 21, 1956, Allum and D. R. Progulske (SC11–56). SDC. Species: 16, 20, 35, 39, 40, 41a, 44, 47, 48, 58, 61, 65, 81, 91.

D: June 18, 1958, Allum and R. B. Kuehne. UMMZ 173827. Species: 42.

STATION 27.—Trib. to Big Sioux R., 6 mi. SSW Brookings, T. 109 N, R. 50 W, sec. 22, SW ¼, Brookings Co., May 8, 1957, Allum and students (SC2–57). Murky pool, 69° F.; bottom sand, gravel, some silt; no vegetation; current slight; depth to 3 feet; width to 30 feet. SDC. Species: 16, 20, 35, 39, 40, 41a, 44, 48, 49, 58, 81, 91, 92.

STATION 28.—Branch of Deer Cr., 2 mi. SE White, T. 111 N, R. 48 W, Brookings Co., May 7, 1952, Allum (FR36–52). Water clear; bottom rock, gravel, sand; vegetation none; depth to 3 feet. SDF. Species: 20, 22, 29, 35, 39, 44, 48, 49, 73, 91.

STATION 29.—Medary Cr., 7 mi. SE Brookings, T. 109 N, R. 49 W, sec. 19, Brookings Co., May 17, 1957, Allum and students (SC4–57). Water clear; bottom sand, gravel; some filamentous algae and *Potamogeton*; current sluggish to moderate; depth to 30 inches; width 10 to 30 feet. SDC. Species: 29, 35, 39, 40, 41a, 44, 47, 48, 49, 58, 91.

STATION 30.—Lake Henry, 6 mi. SW Lake Preston, T. 110 N, R. 55 and 56 W, Kingsbury Co., Aug. 19, 1949, Gibbs and Allen (G11–49). A shallow, weedy duck lake. SDF. Species: 48, 81, 87.

STATION 31.—Brant L., 3 mi. NW Chester, T. 105 N, R. 51 W, Lake Co., Aug. 28, 1949, Gibbs and Allen (G14–49). A large open-water lake; depth about 14 feet; bottom in shallow water rock, gravel, and sand; vegetation sparse. SDF. Species: 20, 35, 48, 58, 61, 78, 81, 87, 91, 92.

STATION 32.—Willow Cr., 3 m. W Sioux Falls, T. 101 N, R. 50 W, Minnehaha Co., Sept. 19, 1939, G. S. Myers and W. A. Gosline (3). UMMZ 127674–75. Species: 40, 42.

STATION 33.—Covell L., Sioux Falls, Minnehaha Co., June 1, 1950, Gibbs and Dahl (G47–50). A small shallow lake subject to winterkill. SDF. Species: 16, 18, 48, 61, 78, 79, 81, 87.

Vermillion River Drainage

STATION 34.—Swan L., 4.5 mi. NW Viborg, T. 97 N, R. 53 W, Turner Co., July 3, 1943, Earl Hansen. UMMZ 163220–30. Species: 18, 20, 40, 42, 44, 48, 61, 78, 81, 87; hybrid 78 × 81.

STATION 35.—Vermillion R., Vermillion, Clay Co., July 5, 1934, C. L. Hubbs and L. C. Hubbs (M34–12). Water visibility few inches only; bottom mud, mostly very soft; scanty emergent marginal vegetation; current slight; depth to 3 feet; width to 65 feet. UMMZ 127396–407. Species: 16, 40, 41a, 42, 48, 51, 54, 61, 71, 73, 78, 81.

James River Drainage

STATION 36.—Richmond L., 8 mi. NW Aberdeen, T. 124 N, R. 65 W, Brown Co., Sept. 27, 1949, Gibbs and Allen (G33–49). An impoundment on Foot Creek. SDF. Species: 48, 61, 77, 80, 91, 92.

STATION 37.—Wiley L., 3 mi. NW Aberdeen, T. 123 N, R. 64 W, Brown Co., Aug. 24, 1951, Gibbs and Allum (FR3–51). A small, shallow municipal pond with much aquatic vegetation. SDF. Species: 14, 18, 20, 58, 61, 77, 78, 79, 80, 84, 87, 92; hybrid 78 × 79.

STATION 38.—James R., 4.3 mi. E Mellette, T. 120 N, R. 63 W, sec. 34, Spink Co., Aug. 25, 1952, Allum and Bailey (B52–60). Water murky, 71° F.; bottom mud and silt; no vegetation; current sluggish; depth to 4 feet; width 50 to 60 feet. UMMZ 166853–62. Species: 16, 41a, 48, 51, 58, 61, 78, 84, 87, 92.

STATION 39.—Thunder (Timber) Cr., U.S. 212, 4 mi. E Frankfort, T. 116 N, R. 62 W, sec. 1, Spink Co., Aug. 14, 1952, Gibbs and Bartel. SDF. Species: 14, 16, 18, 20, 35, 41a, 44, 48, 51, 58, 65, 78, 81, 83, 84, 87, 91, 92.

STATION 40.—Lake Iroquois, 2 mi. SE Iroquois, T. 110 N, R. 58 W, sec. 8, Kingsbury Co., Gibbs and Allen, Aug. 19, 1949 (G12–49). A small artificial impoundment of Marsh Creek. SDF. Species: 18, 77, 78, 81.

STATION 41.—James R., 10 mi. NE Woonsocket, T. 108 N, R. 61 W, Sanborn Co., Jan. 8, 1952, Bartel and Harris (FR24–51). A dipnet collection from an open spring hole during a winterkill. SDF. Species: 20, 41a, 44, 48, 49, 87, 91, 92.

STATION 42.—James R., 9 mi. E and 1 mi. N Woonsocket, T. 107 N, R. 60 W, Sanborn Co., Jan. 8, 1952, Gibbs and Allum (FR25–52). A dipnet collection from an open spring hole during a winterkill. SDF. Species: 44, 49, 58.

STATION 43.—James R., 9 mi. E Woonsocket at bridge, 3 mi. NE Forestburg, T. 107 N, R. 60 W, Sanborn Co., Nov. 13, 1951, Gibbs and Allum (FR18–51). Water turbid, 39° F.; bottom gravel and silt; no vegetation; depth to 3.5 feet; width about 50 feet. UMMZ 163814–25. Species: 16, 20, 40, 41a, 44, 48, 49, 78, 81, 83, 91, 92.

STATION 44.—Sand Cr., 6 mi. E Woonsocket, T. 107 N, R. 61 W, Sanborn Co., Nov. 13, 1951, Gibbs, Allum, and Bartel (FR17–51). Water murky, 39° F.; bottom silty-sand to sand; no vegetation; current slight; depth 3.5 feet; width 30 feet. SDF. Species: 20, 40, 41a, 44, 48, 65, 91, 92.

STATION 45.—Lake Mitchell, just NW Mitchell, Davidson Co., Oct. 25, 1950, Gibbs, Allum, and Peterson (G96–50). A medium-sized impoundment of Firesteel Creek; maximum depth about 40 feet. SDF. Species: 9, 39, 41a.

STATION 46.—James R., 8 mi. ENE Ethan, T. 101 N, R. 58 W, Hanson Co., July 13, 1951, Gibbs, Allum, and Bartel (FR6–51). Water murky; bottom gravel, rubble, silt; collection taken below a mill dam. SDF. Species: 5, 16, 41a, 48, 54, 57, 58, 61, 81, 83, 84, 87.

STATION 47.—Lake Dimock, 3 mi. E Dimock, T. 100 N, R. 60 W, Hutchinson Co., Oct. 30, 1950, Gibbs and Allum (G99–50). A small, shallow impoundment on S Fork of Twelvemile Creek. SDF. Species: 18, 87.

STATION 48.—Pony Cr., 5 mi. ENE Parkston, T. 99 N, R. 60 W, sec. 2, Hutchinson Co., Jan. 5, 1952, Bartel (FR23–52). A dipnet collection from an open spring hole. SDF. Species: 20, 44, 48.

STATION 49.—Dry Cr., 9 mi. ESE Parkston, T. 99 N, R. 59 W, sec. 28, Hutchinson Co., Oct. 12, 1951, Allum and Bartel (FR1–51). An intermittent stream; collection taken in a pool with silt bottom. SDF; UMMZ 163810–11. Species: 16, 39, 41a, 44, 48, 58, 78, 81.

STATION 50.—Lake Henry, Scotland, T. 96 N, R. 58 W, sec. 9, Bon Homme Co., Aug. 29, 1949, Gibbs and Allen (G15–49). A small, shallow impoundment of Dawson Creek. SDF. Species: 83, 84.

STATION 51.—James R., 12 mi. N Yankton at U.S. 81, T. 95 N, R. 55 W, Yankton Co., July 5, 1934, Hubbs and Hubbs (M34–13). Water silty, visibility about one foot; bottom firm to soft mud; no vegetation; depth to 4 feet; width 80 feet. UMMZ 127408–24. Species: 9, 15, 16, 32, 37, 40, 41a, 51, 52, 54, 61, 63, 78, 81, 83, 84, 93; hybrid 83 × 84.

Eastern Tributaries to Missouri River

STATION 52.—Cooper Cr., 0.5 mi. N Springfield at hwy. 37, Bon Homme Co., June 24, 1953, Allum. Water clear; bottom gravel and rubble; current moderate on riffles; depth to 2 feet; width about 20 feet. UMMZ 164910-11. Species: 71, 72.

STATION 53.—Emanuel Cr., 6 mi. W and 6 mi. S Tyndall, T. 94 N, R. 60 W, Bon Homme Co., July 5, 1934, Hubbs and Hubbs (M34-15). Water murky, visibility about 2 feet; bottom soft mud, bedrock, gravel; dense beds of *Ceratophyllum*, some pondweeds, *Lemna*, and algae; current slight; depth to 3 feet in pools; width 1 to 20 feet. UMMZ 127425-37. Species: 16, 20, 29, 39, 40, 41b, 44, 48, 58, 61, 78, 81, 92.

STATION 54.—Tripp L., 7 mi. SW Tripp, T. 97 N, R. 61 W, sec. 20, Hutchinson Co., Aug. 21, 1951, Gibbs, Peterson, and Bartel. A small, shallow impoundment of branch of Choteau Creek. SDF. Species: 77, 80, 84.

STATION 55.—Choteau Cr., 5.5 mi. S and 2.5 mi. W Avon, near junction with Dry Choteau Cr., T. 94 N, R. 62 W, Bon Homme Co., July 5, 1934, Hubbs and Hubbs (M34-16). Water dirty; bottom gravel and mud; a little algae present; current slight between pools; depth to 5 feet; width 2 to 30 feet. UMMZ 127438-46. Species: 9, 40, 41b, 44, 48, 58, 61, 78, 92.

STATION 56.—Choteau Cr., 10 mi. SSW Avon, T. 93 N, R. 62 W, sec. 13, Bon Homme Co., Nov. 15, 1951, Gibbs and Allum (FR19-51). Water slightly turbid, 35° F.; bottom silty sand and rocks; depth to 3.5 feet. SDF; UMMZ 163826-30. Species: 16, 40, 41b, 44, 48, 58, 61, 66.

STATION 57.—Owens Bay of Lake Andes, T. 96 N, R. 64 W, sec. 6, Charles Mix Co. Water clear; bottom sand and organic debris; algae and *Ceratophyllum* abundant; depth to 3 feet; a natural, shallow lake of about 5600 acres.

A: July 8, 1950, Gibbs, Dahl, and Peterson (G54-50). SDF. Species: 61, 69, 77, 92.

B: Nov. 1, 1950, Gibbs and Allum (G101-50). SDF. Species: 48, 61, 69, 92.

C: Aug. 29, 1952, Allum and Bailey (B52-68). UMMZ 166953-56. Species: 48, 69, 77, 92.

STATION 58.—Crow L., 11 mi. SW Wessington Springs, T. 106 N, R. 66 W, Jerauld Co., Sept. 26-27, 1950, Gibbs, Dahl, and Peterson (G90-50 and G92-50). A shallow lake subject to winterkill. SDF. Species: 16, 48, 58, 61, 78, 81, 87.

STATION 59.—Crow Cr., 21 mi. W Wessington Springs, at hwy. 34, T. 107

N, R. 68 W, sec. 9, Buffalo Co., May 1, 1952, Gibbs and Bartel (FR77–52). SDF. Species: 20, 44, 48, 58, 81.

STATION 60.—Elm Cr., trib to Crow Cr., at hwy. 34, 30 mi. W Wessington Springs, T. 107 N, R. 69 W, sec. 18, Buffalo Co., May 1, 1952, Gibbs and Bartel (FR78–52). SDF. Species: 20, 41b, 48, 58.

STATION 61.—Medicine Cr., at hwy. 34, 0.5 mi. above mouth, 13.6 mi. ESE Pierre, T. 110 N, R. 77 W, sec. 22, Hughes Co., Aug. 21, 1952, Allum, Gibbs, and Bailey (B52–48). Water murky, 78° F.; bottom very soft clay, mud, some gravel; no vegetation; current slight; depth to 3 feet; width 3 to 60 feet. UMMZ 166744–58. Species: 15, 16, 20, 40, 41b, 46, 48, 54, 57, 58, 61, 78, 81, 83, 84, 87.

STATION 62.—Hiddenwood L., 4 mi. NE Selby, T. 124 N, R. 76 W, Walworth Co., Sept. 28, 1949, Gibbs and Allen (G37–49). SDF. Species: 58, 77, 81.

STATION 63.—Eureka L., Eureka, T. 127 N, R. 73 W, McPherson Co., Aug. 15, 1951, Allum and Peterson. A shallow lake subject to winterkill. SDF. Species: 77, 80.

Missouri River

STATION 64.—Missouri R. at mouth of Grand R., 2.5 mi. NW Mobridge, T. 19 N, R. 30 E, sec. 31, Corson and Walworth cos., Aug. 24, 1952, Allum, Bailey, and Harris (B52–59). Water turbid, 72° F.; 79° F. in Grand R.; bottom mostly sand, some silt; no vegetation; current moderate to swift; depth to 8 feet; width about 440 yards. Collection taken by swimming a 35 × 6-foot bag seine downstream along sand bars, mostly in water less than 6 feet deep and within 50 yards of shore, for a distance of a mile below mouth of the Grand River. UMMZ 166840–52, 167110, 167114. Species: 2, 3, 15, 16, 24, 27, 28, 46, 54, 63, 66, 72, 85, 86, 93.

STATION 65.—Missouri R., Farm Island State Park, 3 mi. SE Pierre, T. 110 N, R. 79 W, secs. 12–13, Hughes Co., Aug. 21, 1952, Allum, Bailey, Gibbs, *et al.* (B52–49). Water turbid, 73° to 77° F.; bottom sand, soft silt and mud in backwaters; no vegetation; current slight in backwaters to strong in channels; depth to 9 feet; width nearly one-half mile. Collection taken as at station 64, but two bag seines, 35 and 100 feet long, were used. One shore was seined for about one-half mile, mostly in water 6 to 9 feet deep. UMMZ 166759–71, 167116. Species: 2, 15, 16, 24, 28, 45, 46, 54, 63, 66, 78, 83, 85, 86, 87, 93.

STATION 66.—Hipple L., 4 mi. SE Pierre, Farm Island State Park, T. 110 N, R. 78 W, sec. 7, Hughes Co., May 22, 1952, Allum (FR 38–52). This recreation reservoir, a former backwater lagoon of the Missouri River, is now isolated from the river by an earth fill except during extreme floods. Water murky; bottom soft silt; no vegetation; depth to 4 feet. SDF. Species: 24, 40, 41b, 45.

STATION 67.—Missouri R., 4 mi. S Chamberlain, T. 104 N, R. 71 W, Brule Co., Sept. 19, 1939, Myers and Gosline (4). UMMZ 127676–77. Species: 24, 46.

STATION 68.—Fort Randall Reservoir, Platte Creek arm, T. 98 N, R. 69 W, Charles Mix Co., July 25, 1958, James T. Shields and James W. Sprague. A major impoundment of the Missouri River. UMMZ 176677–78. Species: 34.

STATION 69.—Missouri R. at Wheeler Bridge, U.S. 18, T. 96 N, R. 68 W, sec. 1, Charles Mix Co., July 6, 1934, Hubbs and Hubbs (M34–17). Water turbid, visibility about one-fourth inch; bottom soft clay mud near bank, firm sand in current; no vegetation; current slight to swift; seined to a depth of 4 feet and 30 feet from shore; width one-half mile. UMMZ 127447–51. Species: 15, 24, 28, 63, 66.

STATION 70.—Fort Randall Reservoir, at old Wheeler Bridge site, Charles Mix Co. (see station 69), July 21, 1954, Shields. UMMZ 167781. Species: 50.

STATION 71.—Missouri R., tailwater of Fort Randall Dam, T. 95 N, R. 65 W, sec. 8, Gregory Co., Oct. 26, 1954, Shields. UMMZ 167780. Species: 8.

STATION 72.—Missouri R., 1 to 2 mi. below Fort Randall Dam, T. 95 N, R. 65 W, Charles Mix and Gregory cos., Aug. 29, 1952, Allum, Bailey, Gibbs, *et al.* (B52–69). Water turbid, 75°F.; bottom mostly sand in current and soft mud in backwaters, some boulders and chalky bedrock; no vegetation; current moderate to strong; depth to 8 feet; width one-half to three-fourths mile. Collection taken as at station 64, using 35- and 50-foot bag seines of one-fourth inch mesh. Two crews, each with a boat, worked opposite sides of the river from 2:00 to 5:30 P.M. UMMZ 166957–78, 167111, 167115. Species: 2, 3, 5, 15, 16, 24, 27, 28, 40, 45, 46, 50, 51, 54, 61, 63, 66, 77, 78, 81, 83, 84, 85, 86, 93.

STATION 73.—Missouri R., 2 mi. E Niobrara, Knox Co., Nebraska (across river from Bon Homme Co., S. D.), Aug. 4, 1940, Raymond Johnson and P. Romberg (292). Water very turbid, milky white, 83° F.; bottom firm, dark sand; no vegetation; current very strong; depth to 3 feet; seined with a 25-foot bag seine to 100 yards offshore. UMMZ 135336–43. Species: 2, 15, 24, 28, 39, 40, 46, 63.

STATION 74.—Lewis and Clark L., 2 mi. E Springfield, Bon Homme Co., July 2, 1957, Shields. The Gavins Point Dam impounds this major reservoir of the Missouri River. UMMZ 173682. Species: 26.

STATION 75.—Lewis and Clark L., 3 mi. E Springfield, Bon Homme Co., July 3, 1956, Shields. (See Station 74.) UMMZ 172576. Species: 34.

STATION 76.—Missouri R., at and for 5 mi. downstream from Yankton, Yankton Co., Aug. 30, 1952, Allum, Bailey, Gibbs, et al. (B52–70). Water turbid, 75° F.; bottom sand, mud, some fine gravel; no vegetation; current slight to swift; depth to 6 feet; width one-half mile. Collection taken as at station 64 using two crews each operating 35-foot bag seines from 9 A.M. to 1 P.M. A commercial fisherman provided assistance and transportation. UMMZ 166979–91, 167112–13. Species: 2, 3, 5, 15, 16, 24, 27, 28, 45, 46, 48, 54, 58, 63, 64, 85, 93.

STATION 77.—Missouri R., at mouth of James R., below bridge at hwy. 50, 6 mi. E Yankton, Yankton Co., Aug. 27, 1952, S. D. Dept. Game, Fish, and Parks. UMMZ 167160. Species: 4.

STATION 78.—Missouri R., Vermillion, Clay Co., Sept. 9, 1925, A. Hugener. UMMZ 177298. Species: 1.

STATION 79.—Missouri R., 4 mi. N and 1 mi. W Newcastle, Dixon Co., Nebraska (across river from Clay Co., S. D.), Aug. 28, 1941, R. E. and B. B. Johnson (320). Water silty; bottom fine sand and sticky mud; no vegetation; current strong; depth to 6 feet; seined with 25-foot bag seine to 10 feet from shore. UMMZ 135810–15. Species: 2, 24, 27, 28, 45, 66.

STATION 80.—Missouri R., 3 mi. S Burbank, T. 91 N, R. 50 W, Union Co., June 22, 1956, Underhill (56–010). UMMZ 171143. Species: 34.

STATION 81.—Missouri R., 2.5 mi. SW Elk Point, T. 91 N, R. 50 W, Union Co., June 26, 1956, Underhill (56–012). UMMZ 171144. Species: 34.

STATION 82.—Oxbow l. of Missouri R., 2 mi. SW Elk Point, T. 91 N, R. 50 W, Union Co., June 26, 1956, Underhill (56–011). UMMZ 171142. Species: 34.

STATION 83.—Missouri R., 3 mi. E Ponca, Dixon Co., Nebraska (across river from Union Co., S. D.), Aug. 28, 1941, Johnson and Johnson (318). Water muddy, 71° F.; bottom hard sand to silty sand; no vegetation; depth to 6 feet. UMMZ 135794–805. Species: 5, 15, 16, 24, 40, 45, 46, 51, 54, 63, 78, 81.

Western Tributaries to Missouri River south of Cheyenne River

STATION 84.—Antelope Cr. and pond, trib. to Keyapaha R., at Indian School, Rosebud Indian Reservation, T. 38 N, R. 28 W, sec. 3, Todd Co., July 6, 1934, Hubbs and Hubbs (M34–19). Water fairly clear; bottom rather firm, mostly sand and mud; some submerged and emergent vegetation;

depth to 18 inches; width 3 to 8 feet. UMMZ 127458–74. Species: 16, 18, 20, 30, 39, 41b, 44, 48, 49, 58, 61, 71, 77, 78, 84, 92; hybrid 30 × 49.

STATION 85.—Sand Cr., trib. to Antelope Cr., Rosebud Indian Reservation between Okreek and Mission, T. 39 N, R. 27 W, sec. 34, Todd Co., July 6, 1934, Hubbs and Hubbs (M34–18). Water warm, visibility 4 inches; bottom fairly firm clay and sand; some grass in water; depth to 18 inches; width 8 to 15 feet. UMMZ 127452–57. Species: 22, 48, 61, 71, 92; hybrid 21 × 22.

STATION 86.—White Clay Cr., 1 mi. N Pine Ridge at U. S. 18, T. 36 N, R. 45 W, sec. 36, Shannon Co., July 7, 1934, Hubbs and Hubbs (M34–22). Water warm, very silty, visibility about 1 inch; bottom mostly firm sand, some gravel and mud; vegetation almost none; depth to 3 feet; width 5 to 15 feet. UMMZ 127486–93. Species: 20, 30, 41b, 48, 58, 61, 66, 78.

STATION 87.—White R. at [old] U.S. 18, T. 38 N, R. 47 W, sec. 24, Shannon Co.; July 7, 1934, Hubbs and Hubbs (M34–23). Water chalky, warm; visibility about 1 inch; bottom mostly firm sand and gravel, some mud; no vegetation; depth about 18 inches; width 15 feet. UMMZ 127494–99. Species: 24, 30, 41b, 45, 48, 63.

STATION 88.—White R., 6.5 mi. S Kadoka, at hwy. 73, T. 3 S, R. 22 E, sec. 31, Jackson Co., July 20, 1934, L. C. Stuart and H. Young (1). Water warm, very muddy; bottom gravel and mud; vegetation none; no flow; depth to 3 feet; width to 15 feet. UMMZ 120360–69. Species: 24, 27, 30, 41b, 45, 46, 48, 58, 61, 63.

STATION 89.—Fountainprick Cr., 7 mi. S Kadoka (above Wells Ranch), T. 43 N, R. 35 W, sec. 5, Washabaugh Co., Aug. 2, 1934, L. C. Stuart and A. M. Stebler (3). Water cool and clear; bottom mud; algae abundant; depth to 4 feet; width to 6 feet. UMMZ 120372. Species: 48.

STATION 90.—Hay (= Redstone) Cr., 25 mi. SE Kadoka, T. 42 N, R. 34 W, sec. 9, Washabaugh Co., June 29, 1934, Stuart and Stebler (2). Water warm, clear; bottom mud; chara plentiful; depth to 4 feet. UMMZ 120370–71. Species: 48, 78.

STATION 91.—White R., 4 mi. S Belvidere, T. 3 S, R. 24 E, sec. 17, Jackson Co., Aug. 10, 1931, C. E. Burt (48). Water cool, thick and white; bottom white clay mud; vegetation none; current moderate; depth to 2 feet. UMMZ 97832–34. Species: 24, 27, 45.

STATION 92.—West Br. Rosebud Cr., 0.5 mi. above forks, T. 38 N, R. 30 W, sec. 34, Todd Co., July 7, 1934, Hubbs and Hubbs (M34–20). Water cool, very clear; bottom firm to soft sand, some mud and gravel; some beds of white water buttercup; depth to 18 inches; width 5 to 15 feet. UMMZ 127475–78. Species: 10, 20, 30, 48, 58.

STATION 93.—Rosebud L. at Rosebud, T. 38 N, R. 30 W, sec. 34, Todd Co., June 12, 1951, Gibbs (FR4–51). An impoundment on Rosebud Creek. SDF; UMMZ 163812. Species: 19, 20, 30, 48, 49, 61, 78, 79, 80.

STATION 94.—Little White R., at [old] U. S. 18, T. 38 N, R. 30 W, sec. 18, Todd Co., July 7, 1934, Hubbs and Hubbs (M34–21). Water cool, very clear; bottom bedrock, sand, gravel, boulders, some mud; vegetation, considerable white water buttercup and beds of pondweeds; current moderately swift to rapid; depth to 3 feet; width 40 feet. UMMZ 127479–85. Species: 10, 20, 24, 30, 41b, 57, 58, 66.

STATION 95.—Pond at Belvidere, T. 2 S, R. 24 E, sec. 32, Jackson Co., Aug. 10, 1931, Burt (49). Water warm, clear; bottom muck; algae present; depth to 8 feet. UMMZ 97835–37. Species: 48, 61, 79.

STATION 96.—Bad R., Midland, Haakon Co., July 14, 1950, Bailey and Bailey (B50–37). Water murky, 74° F.; bottom sand, gravel, silt; some algae; no flow; depth to 2 feet; width of pools 15 to 25 feet. UMMZ 161918–23. Species: 24, 30, 41b, 45, 48, 81.

Cheyenne River Drainage

STATION 97.—Cheyenne R., T. 8 S, R. 4 E, sec. 36, Fall River Co., Sept. 14, 1950, Gibbs, Dahl, and Peterson (G88–50). SDF. Species: 24, 30, 41b, 45, 48, 61, 78.

STATION 98.—Cascade Cr. at Cascade Falls, T. 8 S, R. 5 E, sec. 30, Fall River Co., Sept. 13, 1950, Gibbs, Dahl, and Peterson (G85–50). SDF. Species: 24, 30, 48, 58, 78.

STATION 99.—Cheyenne R., at confluence with Cascade Cr., T. 8 S, R. 5 E, sec. 31, Fall River Co., Sept. 9, 1949, Gibbs and Kocer (G43–19). SDF. Species: 24, 41b, 45.

STATION 100.—Horsehead Cr., trib. to Angostura Res., T. 9 S, R. 7 E, sec. 27, Fall River Co., Apr. 20, 1952, Gibbs and Bartel (FR75–52). SDF. Species: 18, 45, 48, 58, 61, 77, 78.

STATION 101.—Horsehead arm of Angostura Res., T. 8 S, R. 6 E, sec. 33, Fall River Co., Sept. 13, 1950, Gibbs, Dahl, and Peterson (G86–50). Angostura Reservoir is a large, deep (about 100 feet) impoundment of the Cheyenne River. SDF. Species: 61, 86, 87.

STATION 102.—Cheyenne R., just below Angostura Dam, T. 8 S, R. 6 E, sec. 20, Fall River Co., Nov. 30, 1949, Gibbs and Allen (G44–49). SDF. Species: 24, 30, 45.

STATION 103.—Fall R., just below Hot Springs, T. 7 S, R. 6 E, sec. 30,

Fall River Co., Sept. 12, 1950, Gibbs, Dahl, and Peterson (G84–50). SDF. Species: 11, 30, 60.

STATION 104.—Cheyenne R., road crossing NW of Oral, T. 7 S, R. 7 E, sec. 28, Fall River Co., July 21, 1952, Gibbs and Bartel (FR89–52). SDF. Species: 20, 24, 30, 41b, 45, 48, 54, 58, 70.

STATION 105.—Beaver Cr., 7 mi. NE Hot Springs, T. 6 S, R. 7 E, sec. 15, Custer Co., June 23, 1935, Pierce Brodkorb. UMMZ 108210–11. Species: 30, 58.

STATION 106.—French Cr., 5 mi. E Custer (above new dam of Stockade L.), T. 3 S, R. 5 E, sec. 21, Custer Co., Aug. 27, 1934, Stuart and Crane (6). Water slightly muddy, cool; bottom gravel; algae abundant; current slow; depth to 5 feet. UMMZ 120380–84. Species: 23, 30, 58, 60; hybrid 58 $\times$ 60.

STATION 107.—Iron Cr., trib. to Battle Cr., about 5 mi. S Keystone, T. 2 S, R. 6 E, sec. 27, Custer Co., Aug. 28, 1934, Stuart and Crane (7). Water clear, cool; bottom gravel and boulders; scanty algae; riffles and pools; depth to 4 feet. UMMZ 120385–86. Species: 30, 60.

STATION 108.—Grace Coolidge Cr., Custer State Park, 1 mi. above game lodge, T. 3 S, R. 6 E, sec. 22, Custer Co., Aug. 7, 1934, Stuart and Stebler (4). Water clear, cool; bottom gravel; some algae; current, pools to rapids; depth to 3 feet. UMMZ 120373–76. Species: 30, 58, 60; hybrid 58 $\times$ 60.

STATION 109.—Grace Coolidge Cr., 3.5 mi. above Hermosa at hwy. 36, T. 3 S, R. 7 E, sec. 36, Custer Co., July 8, 1934, Hubbs and Hubbs (M34–24). Water clear, cool; bottom sand, gravel, and boulders, some mud; mostly long pools, some riffles; depth to 2 feet; width 4 to 15 feet. UMMZ 127500–03. Species: 10, 23, 30, 58, 60.

STATION 110.—Battle Cr., 1 mi. SW Hermosa, T. 2 S, R. 8 E, sec. 33, Custer Co., Aug. 11, 1931, Burt. UMMZ 64495. Species: 10.

STATION 111.—Spring Cr., trib. to Cheyenne R., 8 mi. N Hermosa, T. 1 S, R. 8 E, sec. 20, Pennington Co., June 14, 1929, Burt (1). UMMZ 87301–05. Species: 20, 23, 48, 58.

STATION 112.—Castle Cr., 5 mi. above Deerfield Reservoir, T. 1 N, R. 2 E, sec. 16, Pennington Co., Aug. 16, 1950, Gibbs, Dahl, and Peterson (G79–50). SDF. Species: 10, 12, 30, 60.

STATION 113.—Castle Cr., behind Deerfield store, T. 1 N, R. 2 E, sec. 26, Pennington Co., Aug. 15, 1950, Gibbs, Dahl, and Peterson (G76–50). SDF. Species: 30, 58, 60.

STATION 114.—S Fk. Castle Cr., 1 mi. above confluence with Castle Cr., T. 1 N, R. 2 E, sec. 36, Pennington Co., Aug. 15, 1950, Gibbs, Dahl, and Peterson (G77–50). SDF. Species: 23, 30, 60.

STATION 115.—Rapid Cr., Placerville [= 0.5 mi. E Pactola Dam], T. 1 N, R. 5 E, sec. 2, Pennington Co., Aug. 11–28, 1938, J. Breukelman (126–129). Water clear, 63° F.; bottom rocky; pondweeds present; depth to 3 feet; width 10 feet. UMMZ 126924–33. Species: 23, 30, 58, 60.

STATION 116.—Rapid Cr., 2 mi. below Rapid City, T. 1 N, R. 8 E, sec. 8, Pennington Co., July 8, 1934, Hubbs and Hubbs (M34–25). Water warm, quite clear, bottom visibility 3 feet; bottom sand, mud, gravel, boulders; dense growth of pondweeds; current moderate; depth greater than 4 feet; width 25 to 50 feet. UMMZ 127504–08. Species: 10, 20, 30, 41b, 58, 60.

STATION 117.—Rapid Cr., 2 mi. W Farmingdale, T. 1 S, R. 10 E, sec. 15, Pennington Co., July 17, 1930, A. and R. D. Svihla. Bottom mud and gravel; depth to 4 feet. UMMZ 92336–39. Species: 24, 30, 41b, 57.

STATION 118.—Rapid Cr., 4 mi. SE Caputa, T. 1 S, R. 10 E, sec. 16, Pennington Co., Aug. 11, 1931, Burt (50). Water green and fairly clear, warm; bottom rock and gravel; weeds and willows along bank; current moderate; depth to 4 feet. UMMZ 97838–42. Species: 20, 24, 41b, 48, 57.

STATION 119.—Box Elder Cr., 4 mi. N (NW ?) Nemo, T. 3 N, R. 5 E, sec. 18, Lawrence Co., July 16, 1930, Svihla and Svihla. UMMZ 92340. Species: 10.

STATION 120.—Cheyenne R., 2.5 mi. E Wasta, at U. S. 14 and 16, T. 1 N, R. 14 E, sec. 2, Pennington Co. Water turbid; bottom gravel and rubble on riffles, soft silt overlying sand elsewhere; no vegetation; current slight to swift; depth to 3 feet; width 60 to 300 feet.

A: Sept. 20, 1939, Myers and Gosline (5). UMMZ 127678. Species: 24.

B: July 14, 1950, Bailey and Bailey (B50–36). UMMZ 161905–17. Species: 24, 27, 30, 41b, 45, 54, 57, 58, 61, 63, 66, 78, 81.

C: Nov. 29, 1951, Gibbs and Bartel (FR20–51). UMMZ 163831–34. Species: 27, 41b, 63, 70.

STATION 121. Mirror L., trib. to Crow [= Beaver] Cr., 9 mi. NW Spearfish, T. 7 N, R. 1 E, sec. 20, Lawrence Co., Aug. 22, 1952, Allum, Bailey, and Gibbs (B52–52). Water clear, easily roiled, cold (60° F.; air, 92° F.); bottom soft red mud and marl; vegetation common to abundant; depth of water to 25 feet, of capture to 3 feet; area about 3 acres. This lake supports rainbow and brown trout although none was caught. UMMZ 166776–80. Species: 21, 30, 41b, 48, 58.

STATION 122.—Mud L., 8 mi. NW Spearfish, T. 7 N, R. 1 E, sec. 16, SE ¼, Lawrence Co., Aug. 22, 1952, Allum, Bailey, and Gibbs (B52–50). Water crystal clear, 76° to 80° F., with sulfur odor; bottom marl; chara abundant over bottom; depth rather uniformly about 30 inches; area about 2 acres. UMMZ 166772–73. Species: 21, 78.

STATION 123.—Redwater Cr., at mouth of Crow (or Beaver) Cr., 9 mi. W Spearfish, T. 7 N, R. 1 E, sec. 16, NW ¼, Lawrence Co., Aug. 22, 1952, Allum, Bailey, and Gibbs (B52–53). Water clear, 68° F.; bottom mud, rubble, gravel; some water cress, *Potamogeton,* and algae; current moderate to swift; depth to 7 feet. UMMZ 166781–85. Species: 21, 30, 58, 59, 60.

STATION 124.—Cox L., 8 mi. NW Spearfish, T. 7 N, R. 1 E, sec. 16, SE ¼, Lawrence Co., Aug. 22, 1952, Allum, Bailey, and Gibbs (B52–51). Water crystal clear, sulfur odor, 54° F. in open water, 74° at outlet (air, 92° F.); bottom marl, easily roiled; algae abundant, some chara; depth reportedly between 63 feet and 82 feet; area perhaps 2 acres. This lake, really a gigantic spring, supports trout; the outlet, Lake Creek, contains trout in the spring. UMMZ 166774–75. Species: 21, 30.

STATION 125.—Spearfish Cr., 2 mi. NNW Spearfish, at junction hwys. 14 and 85, Lawrence Co., July 13, 1950, R. M. Bailey and D. M. Bailey (B50–35). Water clear, 68° F.; bottom gravel and rubble; moderate algae; current moderate; depth to 3 feet; width 12 to 18 feet. UMMZ 161904. Species 60.

STATION 126.—Spearfish Cr., T. 7 N, R. 2 E, sec. 16, Lawrence Co., Sept. 25, 1952, Allum. UMMZ 163808–09. Species: 59, 60.

STATION 127.—Redwater R., near confluence with Belle Fourche R., just E Belle Fourche at U.S. 212, T. 8 N, R. 2 E, sec. 11, Butte Co., Nov. 27, 1951, Gibbs and Bartel (FR21–51). SDF. Species: 16, 20, 30, 41b, 45, 48, 58, 61, 78.

STATION 128.—Belle Fourche Res. (Orman L.), near S end of dam, T. 9 N, R. 4 E, sec. 19, Butte Co., Aug. 22–23, 1952, Allum, Bailey, and Gibbs (B52–54). Water murky near shore, 72° F.; bottom gravel and soft silt; no rooted vegetation; depth of capture to 15 feet; height of dam 112 feet; reservoir capacity 177,510 acre feet; collection by gill net and seines. UMMZ 166787–95. Species: 16, 24, 41b, 45, 54, 57, 58, 59, 63, 78, 86, 87.

Western Tributaries to Missouri River north of Cheyenne River

STATION 129.—Moreau R., at hwy. 65, 11.5 mi. NNE Dupree, T. 14 N, R. 21 E, sec. 2, Ziebach Co., Aug. 24, 1952, Gibbs and Bartel (B52–66). Water slightly turbid, 78° F.; bottom gravel, sand, silt; some bulrushes; current rapid; depth to 5 feet. UMMZ 166929–40. Species: 24, 30, 41b, 45, 48, 54, 57, 58, 61, 63, 66, 78.

STATION 130.—Moreau R., at hwy. 63, 14 mi. N Eagle Butte, T. 14 N, R. 24 E, secs. 7 and 8, Dewey Co., Aug. 24, 1952, Gibbs and Bartel (B52–67). Water very turbid, 80° F.; bottom gravel, sand, silt; some bulrushes;

current swift; depth to 4 feet. UMMZ 166941–52. Species: 15, 24, 30, 41b, 45, 46, 48, 54, 61, 63, 66, 78, 87.

STATION 131.—S Fk. Grand R., Buffalo, at U. S. 85, T. 19 N, R. 5 E, sec. 29, Harding Co., Aug. 23, 1952, Allum, Bailey, and Harris (B52–56). Water turbid from recent rain, 78° F.; bottom sand, some silt; rushes near shore; current moderate; depth to 12 inches. UMMZ 166811–16. Species: 24, 30, 41b, 45, 48, 58.

STATION 132.—S Fk. Grand R., 6 mi. N Bison, T. 19 N, R. 13 E, sec. 13, Perkins Co., Aug. 23, 1952, Allum, Bailey, and Harris (B52–57). Water turbid, 74° F.; bottom sand, gravel, rubble, silt; emergent grass; current moderate; depth to 2 feet; width about 50 feet. UMMZ 166817–27. Species: 20, 24, 30, 41b, 45, 48, 54, 57, 58, 61, 66.

STATION 133.—Flat Cr., at upper end of Flat Creek L., T. 21 N, R. 16 E, sec. 18, Perkins Co., June 7, 1950, Gibbs, Dahl, and Peterson (G50–50). SDF. Species: 45, 48, 58, 78, 79, 81; hybrid 78 $\times$ 81.

STATION 134.—Flat Creek L., 10 mi. S Lemmon at hwy. 73, T. 21 N, R. 16 E, sec. 18, Perkins Co., Sept. 29, 1949, Gibbs and Allen (G39–49). Flat Creek Lake is a small impoundment of Flat Creek. SDF. Species: 47, 77, 78, 80, 81, 84, 87.

STATION 135.—Grand R. at U. S. 12, 14 mi. NW Mobridge, T. 20 N, R. 28 E, sec. 26, Corson Co., Aug. 24, 1952, Allum, Bailey, and Harris (B52–58). Water very turbid, 82° F.; bottom sand, some boulders and silt at bridge; no vegetation; current moderate; depth to 9 feet, mostly less than 18 inches; width 40 to 70 feet. UMMZ 166828–39. Species: 15, 16, 24, 27, 30, 46, 48, 54, 57, 63, 66, 85, 86.

Little Missouri River Drainage

STATION 136.—Little Missouri R., at hwy. 8, Camp Crook, T. 18 N, R. 1 E, sec. 2, Harding Co., Aug. 23, 1952, Allum, Bailey, and Gibbs (B52–55). Water very turbid from recent rain, 83° F.; bottom gravel and silt; no vegetation; current moderate; depth to 4 feet, mostly 12 inches; width 20 to 50 feet. UMMZ 166796–810. Species: 15, 16, 23, 24, 30, 41b, 44, 45, 46, 48, 54, 57, 58, 61, 63, 66.

STATION 137.—Box Elder Cr., 28 mi. N Camp Crook, T. 23 N, R. 1 E, sec. 22, Harding Co., Aug. 23, 1952, Gibbs and Bartel (B52–65). Water slightly turbid, 81° F.; bottom gravel, sand, silt; some bulrushes and narrow-leafed *Potamogeton;* current slow; depth to 4 feet. UMMZ 166921–28. Species: 16, 24, 30, 41b, 45, 48, 58, 66.

ANNOTATED LIST OF FISHES

PETROMYZONTIDAE—LAMPREYS

KEY TO LAMPREYS

1a.—Circumoral teeth unicuspid (occasionally 1 or a few bicuspid). Transverse lingual lamina usually moderately to strongly bilobed. Supraoral cusps 1 to 4 (usually 1 or 2). Teeth in lateral rows 5 to 8, usually 6 or 7. Teeth in anterior row 2 to 4, usually 3. .... Silver lamprey, *Ichthyomyzon unicuspis*
1b.—Circumoral teeth in part (1 to 11, usually 6 to 8) bicuspid. Transverse lingual lamina usually linear or weakly bilobed. Supraoral cusps 2 or 3. Teeth in lateral rows 6 to 11, usually 8 or 9. Teeth in anterior row 3 to 5, usually 4 or 5. (*Hypothetical in South Dakota.*) .... Chestnut lamprey, *Ichthyomyzon castaneus*

1. *Ichthyomyzon unicuspis* Hubbs and Trautman—Silver lamprey

(?) *Ichthyomyzon concolor.* Evermann and Cox, 1896: 384 (Crow Cr., [near] Chamberlain).
Ichthyomyzon unicuspis. Bailey, 1959b: 162–63 (description; Missouri R., Vermillion). Underhill, 1959: 99 (Missouri R., at mouth Vermillion R.*).

The small specimen reported by Evermann and Cox is assigned here on a presumptive basis; no characters were reported and the specimen is not known to be extant.

Station record: 78.

ACIPENSERIDAE—STURGEONS

KEY TO STURGEONS

1a.—Caudal peduncle incompletely armored, short and compressed. Snout narrower and deeper, more or less blunt and rounded in adults. Spiracle and pseudobranchium present. Barbels not fringed. Caudal fin without filament. (*Hypothetical in South Dakota.*) .... Lake sturgeon, *Acipenser fulvescens*
1b.—Caudal peduncle completely armored, long and much depressed. Snout greatly expanded and depressed, shovellike. No spiracle or pseudobranchium. Barbels coarsely fringed. Upper lobe of caudal produced into an elongate filament (often injured in adults) .... *Scaphirhynchus* 2
2a.—Belly covered with a mosaic of dermal plates (except in young). Bases of outer barbels in a line with or ahead of inner barbels. Inner barbel heavily fringed and long. All barbels placed farther forward on snout. Dorsal fin rays 30 to 36; anal fin rays 18 to 23. Lateral plates larger; eye larger; snout blunter; color darker. Size smaller, maximum weight about 5 pounds, usually much less ....Shovelnose sturgeon, *Scaphirhynchus platorynchus*
2b.—Belly largely naked at all ages. Bases of outer barbels lying behind inner barbels. Inner barbel weakly fringed and short. All barbels placed farther back on snout. Dorsal rays 37 to 43; anal rays 24 to 28. Lateral plates smaller; eye smaller; snout sharper; color more pallid. Size larger, maximum weight over 60 pounds .... Pallid sturgeon, *Scaphirhynchus albus*

* Specimen (s) examined by authors.

2. *Scaphirhynchus platorynchus* (Rafinesque)—Shovelnose sturgeon

Scaphirhynchus platorhynchus. Meek, 1892: 245 (Missouri R., Sioux City, Iowa [and S. D.]).
Scaphirhynchus platorynchus. Churchill and Over, 1933: 19–20, fig. 3 (description, ecology, importance; Missouri R.). Bailey and Cross, 1954: 169–99, fig. 10 (synonymy, description, habitat, distribution; Missouri R., near mouth of Grand R., Corson Co.*; Missouri R., 3 mi. SE Pierre, Hughes Co.*; Missouri R., 1–2 mi. below Ft. Randall Dam, Gregory Co.*; Missouri R., [near] Yankton, Yankton Co.*). Shields, 1958b: 360 (Ft. Randall Res.). Underhill, 1959: 99 (mouth of Vermillion R.).

This common species is apparently restricted in South Dakota to the Missouri River and the lower reaches of its larger tributaries.

Station records: 64, 65, 72, 73, 76, 79.

3. *Scaphirhynchus albus* (Forbes and Richardson)—Pallid sturgeon

Acipenser rubicundus (misidentifications). Evermann and Cox, 1896: 385 (hearsay reports from mouth of White R. and Missouri R. near Chamberlain, and Missouri R. near Yankton).
Scaphirhynchus album. Bailey and Cross, 1954: 169–190, 199–202, fig. 10 (synonymy, description, habitat, distribution; Missouri R., just below mouth of Grand R., Corson Co.*; Missouri R., 2 mi. below Ft. Randall Dam, Gregory Co.*; Missouri R., [near] Yankton, Yankton Co.*).

The pallid sturgeon is much less common than the shovelnose sturgeon, but like that species it is confined to the Missouri River and the lower parts of major tributaries. An account and photo of an individual taken in the Missouri River near Washburn, North Dakota (North Dakota Outdoors, July, 1956) indicate that this sturgeon reaches a weight of at least 68 pounds.

The International Code of Zoological Nomenclature (1961, art. 30) specifies that generic names ending in *-rhynchus* should be treated as of masculine gender. The adjectival specific name of the pallid sturgeon is therefore emended to *albus*.

Station records: 64, 72, 76.

POLYODONTIDAE—PADDLEFISHES

4. *Polyodon spathula* (Walbaum)—Paddlefish

Polyodon folium. Girard, 1858: 358 (description; [Missouri R.] Ft. Pierre, "Nebraska").
Polyodon spathula. Meek, 1892: 245 (Missouri R., Sioux City, Iowa [and S. D.]). Evermann and Cox, 1896: 385 (description; mouth of White R., [near] Chamberlain). Churchill and Over, 1933: 19, fig. 2 (description, ecology, importance; Missouri R. and larger tributaries). Cleary, 1956: 276 (Big Sioux R., Lyon [Lincoln] and Sioux [Union] cos., Iowa [and S. D.]). Hugghins, 1959: 21 (Missouri R., Ft. Randall Res.). Underhill, 1959: 99 (mouth of Vermillion R.).

Paddlefish have been reported from the Missouri River and from Fort Peck Reservoir as well as from the Milk River at Glasgow, all in Montana (Brown, 1951: 252), so it is evident that the species occurs throughout the South Dakota section of the Missouri River. It has lately supported an intensive winter sport fishery in the tailwaters of Fort Randall Dam (4491 fish taken during the period, November 15, 1958 to February 28, 1959, according to data assembled by James W. Sprague). The fishery at Gavin Point Dam is much smaller. Except in the Missouri River, the paddlefish has been reported in South Dakota only from the Big Sioux River (Cleary, 1956) and the mouths of other tributaries, but we have heard rumors of an occasional capture well upstream in the James River.

Station record: 77.

LEPISOSTEIDAE—GARS

KEY TO GARS

1a.—Snout short and broad, its least width contained about 5 to 7 times in its length (except in young). Interorbital width about 1.7 in postorbital length of head. Scale rows around caudal peduncle 26 to 30.........Shortnose gar, *Lepisosteus platostomus*

1b.—Snout long and narrow, its least width contained about 12 to 20 times in its length (except in young). Interorbital width usually about 2.0 in postorbital length of head. Scale rows around caudal peduncle 19 to 24 ....Longnose gar, *Lepisosteus osseus*

5. *Lepisosteus platostomus* Rafinesque—Shortnose gar

Lepidosteus productus (misidentification). Cope, 1879: 441 (characters; Missouri R. pools, near Battle Cr. [? = Blue Blanket Cr.]).
Cylindrosteus platostomus. Churchill and Over, 1933: 21, fig. 5 (description, ecology, importance; larger streams and many lakes).
Lepisosteus platostomus. Shields, 1958a: 31–32; 1958b: 360 (Fort Randall Reservoir). Underhill, 1959: 99 (lower ten miles of Vermillion R.).

This is much the more common of the two gars in South Dakota. Since it occurs upstream as far as Fort Peck Dam, Montana (specimens examined by Bailey in Montana State University collection), it may be expected throughout the South Dakota section of the Missouri River. It has been taken elsewhere in the state only in the lower parts of the James and Vermillion rivers.

Station records: 46, 72, 76, 83.

6. *Lepisosteus osseus* (Linnaeus) —Longnose gar

Lepidosteus otarius. Cope, 1879: 441 (characters; pools of Missouri R. near Battle Cr. [? = Blue Blanket Cr.]).
Lepisosteus osseus. Evermann and Cox, 1896: 386 (description; Crow Cr. [near] Chamberlain). Churchill and Over, 1933: 20–21, fig. 4 (description, ecology, importance;

Missouri and larger rivers and Big Stone L.). Cleary, 1956: 278 (Big Sioux R., Sioux and Woodbury cos., Iowa [and Union Co., S. D.]). Underhill, 1959: 99 (mouth of Vermillion R.).

This rather uncommon South Dakota species is not represented in our collections, but an adult from Gavins Point Reservoir has been identified by us. It has been reported from the lower part of some large Missouri River tributaries, and from Big Stone Lake.

AMIIDAE—BOWFINS

7. *Amia calva* Linnaeus—Bowfin

Amia calva. Churchill and Over, 1933: 21–22, fig. 6 (description, ecology, importance; ". . . found rather generally in the streams and lakes of eastern South Dakota.").

Despite the above-cited statement by Churchill and Over, we have not encountered this species in South Dakota. It is common and widely distributed in southern and central Minnesota (Eddy and Surber, 1947: 85), so it presumably occurs in Big Stone Lake. It is possible that bowfins were formerly present and were almost or quite extirpated from South Dakota during the drought period of 1933 to 1939. However, we know of no firm evidence of the natural occurrence of the bowfin in the Missouri River drainage. It has recently been introduced in the Okoboji Lake area of northwestern Iowa (Cleary, 1956: 278), and Johnson (MS) heard reports of its presence in fish hatchery ponds in Nebraska and of its introduction from Minnesota.

CLUPEIDAE—HERRINGS

KEY TO HERRINGS

1a.—Mouth terminal, the lower jaw protruding well beyond upper. Maxilla extends to below center of eye. Dorsal-fin origin in front of pelvic insertion. Posterior ray of dorsal fin not prolonged into a filament. Gill rakers few, about 22 on lower limb of first arch .... Skipjack herring, *Alosa chrysochloris*
1b.—Mouth subterminal, the lower jaw shorter. Maxilla extends to below front of eye. Dorsal origin behind pelvic insertion. Posterior ray of dorsal fin prolonged into a prominent filament (except in tiny young). Gill rakers numerous ....
.... Gizzard shad, *Dorosoma cepedianum*

8. *Alosa chrysochloris* (Rafinesque)—Skipjack herring

Pomolobus chrysochloris. Eddy and Surber, 1943: 90–91 (description, habitat; ". . . at one time . . . common in Big Stone Lake").

The skipjack herring formerly occurred in Big Stone Lake, but it is now probably extirpated from the upper Mississippi basin (p. 121). One individual was taken at Fort Randall Dam on the Missouri River.

Station record: 71.

9. *Dorosoma cepedianum* (LeSueur)—Gizzard shad

Dorosoma cepedianum. Meek, 1892: 245–46 (Missouri R. and Big Sioux R.*, Sioux City, Iowa [and S. D.]). Churchill and Over, 1933: 23–24, fig. 8 (description, ecology, importance; Missouri R. and eastern tributaries). Cleary, 1956: 280 (Big Sioux R., Lyon [Lincoln], Sioux [Lincoln and Union], and Plymouth [Union] cos., Iowa [and S. D.]). Shields, 1958a: 31; 1959b: 360 (Fort Randall Res.). Underhill, 1959: 99 (Vermillion R., from Centerville to mouth).

Gizzard shad live in the Missouri River and tributaries as far upstream as Fort Randall Reservoir, but they have not been taken in Oahe Reservoir. The species occurs in the Minnesota River in Minnesota (Eddy and Surber, 1947: 91), but has not been taken in that drainage in South Dakota.

Station records: 45, 51, 55.

SALMONIDAE—TROUTS, WHITEFISHES, AND GRAYLINGS

KEY TO TROUTS

1a.—Scales larger, fewer than 140 in series just above lateral line. Body and fins with more or less definite dark spots. Prevomer flattened, the shaft itself bearing 1 or 2 rows of teeth (these not on a free crest), the posterior teeth often lost with age. Parr-marks (when evident, especially in young) scarcely or not wider than interspaces .. *Salmo* 2

2a.—Dark spots larger, fewer and more irregular; faint or absent on caudal. Adipose fin with a light margin, more or less orange in life (especially in young). Orange or reddish spots often present on body. Principal anal rays (including one unbranched anterior ray) typically 9. Dorsal origin much closer to tip of snout than to base of caudal fin (insertion of pelvic below posterior half of dorsal base) .... Brown trout, *Salmo trutta*

2b.—Dark spots numerous, smaller, and sharper; especially marked on caudal. Adipose fin light with a dark margin; often heavily spotted in adults. No orange or reddish spots on body; adults with a broad pink or reddish stripe along side. Principal anal rays 10 to 12 (occasionally 9 in young in which one ray has not yet become branched). Dorsal origin usually about equidistant from base of caudal and tip of snout in young and juveniles, somewhat closer to snout in adults (the insertion of pelvic below anterior half of dorsal base) ......Rainbow trout, *Salmo gairdneri*

1b.—Scales smaller, more than 190 just above lateral line. Body frequently mottled or vermiculated with dark, but without definite small dark spots (red and blue spots often present). Prevomer boat-shaped; the shaft depressed, toothless. Parr-marks (when evident) conspicuously broader than interspaces .... Brook trout, *Salvelinus fontinalis*

10. *Salmo trutta* Linnaeus—Brown trout

Salmo levenensis. Churchill and Over, 1933: 25–26, fig. 10 (description, ecology, importance; streams of Black Hills).
Salmo trutta. Hugghins, 1959: 21 (Black Hills streams).

This European species is stocked widely in the Black Hills and occasionally elsewhere, including Shadehill Reservoir.
Station records: 92, 94, 109, 110, 112, 116, 119.

11. *Salmo gairdneri* Richardson—Rainbow trout

Salmo shasta. Churchill and Over, 1933: 26, fig. 11 (description, ecology, importance; streams of the Black Hills).
Samo gairdneri. Hugghins, 1959: 21 (Black Hills streams).

The rainbow trout is the most widely introduced of the three trouts in South Dakota. In addition to the waters of the Black Hills area, plantings have been made in a number of stock ponds, especially in the western part of the state, and in a few suitable streams and ponds of the Minnesota and Big Sioux river drainages and perhaps elsewhere. Recent stockings include Fort Randall and Lewis and Clark tailwaters on the Missouri River.
Station record: 103.

12. *Salvelinus fontinalis* (Mitchill)—Brook trout

Salvelinus fontinalis. Evermann and Cox, 1896: 415 (Spearfish Cr., Spearfish). Churchill and Over, 1933: 24–25, fig. 9 (description, ecology, importance; streams of the Black Hills). Hugghins, 1959: 21 (Black Hills streams).

The brook trout is stocked principally in the waters of the Black Hills. Only a few suitable ponds and streams outside this area are known to have had introductions.
Station record: 112.

UMBRIDAE—MUDMINNOWS

13. *Umbra limi* (Kirtland)—Central mudminnow

We know of no previous report of this species from South Dakota. The single locality is a small spring run in a pasture 6 miles NE of Brookings. This spring harbors, in addition, a relict population of *Chrosomus eos*. Four mudminnows, a large adult and three yearlings, were collected in January, 1947. Three further visits to the station have failed to yield additional specimens of *Umbra*.
Station record: 25A.

ESOCIDAE—PIKES

14. *Esox lucius* Linnaeus—Northern pike

Lucius lucius. Evermann and Cox, 1896: 415 (Rock Cr., Mitchell). Woolman, 1896: 348, 352, 357 (Lake Traverse; Daugherty Cr. and Little Minnesota R., Browns Valley, Minn. [and S. D.]; Big Stone L., Creagers farm and Ortonville, Minn. [and S. D.]; and Wheatstone [Whetstone] Cr., Milbank).

Esox lucius. Churchill and Over, 1933: 61–62, fig. 49 (description, ecology, importance; lakes and larger streams of NE quarter of state). Hugghins, 1959: 21 (Big Stone L., Waubay L., Willow L., L. Madison, Crow L., L. Louise, Cottonwood L., Big Sioux R., College [= Six-mile] Cr.). Moyle and Clothier, 1959: 178 (Lake Traverse). Underhill, 1959: 99 (introduced; Lake Marindahl [Yankton Co.]).

Northern pike were native to the Red, Minnesota, Big Sioux, and James river drainages, and have been introduced widely throughout South Dakota. Station records: 3, 8, 25B, 37, 39.

HIODONTIDAE—MOONEYES

KEY TO MOONEYES

1a.—Dorsal base about ½ anal base. Dorsal with 11 or 12 principal rays, its origin before anal. Fleshy midventral keel not extending in front of pelvic base. Eye larger, the iris silvery. *(Hypothetical in South Dakota.)* .... Mooneye, *Hiodon tergisus*

1b.—Dorsal base about ⅓ anal base. Dorsal with 9 or 10 principal rays, its origin behind anal. A fleshy keel extending along midventral line from just behind pectorals to vent. Eye smaller, the iris golden .... Goldeye, *Hiodon alosoides*

15. *Hiodon alosoides* (Rafinesque)—Goldeye

Hiodon tergisus (misidentification). Cope, 1879: 441 (Missouri R. pools, near Battle [? = Blue Blanket] Cr.). Churchill and Over, 1933: 23 (compiled, based on Evermann and Cox report [1896: 413] of Cope's record).

Hiodon alosoides. Meek, 1892: 245–46 (Big Sioux and Missouri rivers, Sioux City, Iowa [and S. D.]). Evermann and Cox, 1896: 412 (habitat, size; Crow Cr., [near] Chamberlain; Choteau Cr., Springfield). Cleary, 1956: 281 (Big Sioux R., Sioux [Lincoln] and Woodbury [Union] cos., Iowa [and S. D.]). Underhill, 1959:99 (mouth of Vermillion R.).

Amphiodon alosoides. Churchill and Over, 1933: 22–23, fig. 7 (description, ecology, importance; Cheyenne R. at mouth of Cherry Cr., in James and Vermillion rivers, and other streams, especially west of Missouri R.). Shields, 1958a: 31; 1958b: 360 (Fort Randall Res.).

The goldeye abounds in the Missouri River and is present in moderate numbers in the larger tributaries. In the early years after impoundment of Fort Randall and Gavins Point reservoirs, populations remained at high levels. The goldeye has not been taken in the Minnesota River drainage of South Dakota.

Station records: 51, 61, 64, 65, 69, 72, 73, 76, 83, 130, 135, 136.

CYPRINIDAE—MINNOWS

KEY TO GENERA OF CYPRINIDAE

1a.—Dorsal and anal each with a strong serrated spine; dorsal fin long, with more than 15 soft rays 2

    2a.—Upper jaw with 2 long, fleshy barbels on each side. Lateral-line scales 35 to 38 (body sometimes scaleless—the "leather carp," or partially scaled—the "mirror carp"). Gill rakers on anterior arch 21 to 27. Pharyngeal teeth in 2 rows, 2,3–3,2; those of the main row molariform .... *Cyprinus*

    2b.—Upper jaw without barbels. Lateral-line scales 26 to 29. Gill rakers on anterior arch 37 to 43. Pharyngeal teeth in a single row, 4–4, not molarlike ...... *Carassius*

1b.—No spinous rays in dorsal or anal fins; dorsal fin short, with fewer than 10 developed rays 3

    3a.—Abdomen behind pelvic fins with a fleshy keel over which the scales do not pass. Anal rays 10 to 14, usually 11 to 13. Lateral line greatly decurved. Anal fin falcate. [Teeth usually 5–5.] .... *Notemigonus*

    3b.—Abdomen behind pelvic fins rounded over and usually fully scaled (almost or quite naked in some species of *Hybopsis*). Anal rays 12 or fewer (9 or fewer in most species). Lateral line little decurved. Anal fin infrequently falcate 4

        4a.—Pharyngeal teeth in main row typically 5–5 or 5–4 (4–4 only in rare variants).. 5

            5a.—Maxilla with a flaplike barbel that is placed in a groove above upper lip well in advance of angle of mouth (barbel small or obsolete in young; the mouth should be opened to expose the groove in searching for the barbel). Lateral line complete. Peritoneum silvery .... *Semotilus*

            5b.—No maxillary barbel. Lateral line incomplete. Peritoneum black .. *Chrosomus*

        4b.—Pharyngeal teeth in main row 4–4. 6

            6a.—Maxilla with a slender barbel at its posterior end 7

                7a.—Scale radii restricted to the posterior (exposed) field. Upper jaw protractile, separated from snout by a groove. .... *Hybopsis*

                7b.—Scales with radii in all fields. Upper jaw not protractile, not separated from snout by a groove .... *Rhinichthys*

            6b.—Maxilla without a barbel (a transitory fleshy flap that simulates a barbel is present at the posterior angle of the mouth in breeding males of *Pimephales notatus*) 8

                8a.—Lower lip thick, rugose, with a fleshy projection on each side that is partly separated from mandible by a groove .... *Phenacobius*

                8b.—Lower lip rather thin and smooth, without fleshy lateral projections 9

                    9a.—Cartilaginous ridge of lower jaw, if present, less prominent, and not separated by a definite groove from lower lip. Intestine not spirally looped around the swimbladder. Gill rakers on first arch fewer than 15, rather short 10

                      10a.—Predorsal scales usually neither greatly crowded nor conspicuously smaller than those on rest of body, in 21 or (usually) fewer rows (except in *N. cornutus* which has 9 or more anal rays). Second (rudimentary) ray of dorsal slender and adhering closely to first principal ray. Nuptial organs not confined to a cluster of heavy tubercles on front of head.. 11

                        11a.—Intestine short, much less than twice standard length, with a single S-shaped loop. Peritoneum usually silvery, often flecked with dark (occasionally or regularly black in a few species). Mouth usually terminal, U-shaped, the lower jaw not thin. Suborbitals narrow, 40% or less of cheek (sometimes nearly 50% in *N. atherinoides*) ....*Notropis*

11b.—Intestine elongate, more than twice standard length, with several loops. Peritoneum black. Mouth gently curved, crescent shaped; the lower jaw very thin. Suborbitals broad, extending 40–70% across cheek .... *Hybognathus*

10b.—Predorsal scales crowded, much smaller than those on rest of body, in 21 or more rows. Anal rays 7. Second (rudimentary) ray of dorsal short and stout, separated from first principal ray by a membrane (best developed in adult males). Nuptial tubercles large, those of head and body confined to a cluster on front of snout and (in *P. promelas*) chin .... *Pimephales*

9b.—Cartilaginous ridge of lower jaw prominent and separated by a groove from the fleshy lower lip. Intestine spirally looped about the swimbladder. Gill rakers on first arch 29 to 34, moderately long and slender—*Campostoma*

16. *Cyprinus carpio* Linnaeus—Carp

Cyprinus carpio. Churchill and Over, 1933: 35–37, fig. 22 (description, ecology, importance; absent from some lakes, few west of Missouri R.). Cleary, 1956: 290 (Big Sioux R., Lyon [Lincoln], Sioux [Lincoln and Union], Plymouth [Union], and Woodbury [Union] cos., Iowa [and S. D.]). Shields, 1958a: 30–32; 1958b: 360 (Gavins Point and Fort Randall reservoirs). Allum and Hugghins, 1959: 34 (Brant L., Lake Co.). Hugghins, 1959: 21 (Waubay L., L. Poinsett, Oakwood L., Brant L., L. Chapelle, Cottonwood L., Mina L. [= L. Parmley], Big Sioux R.). Moyle and Clothier, 1959: 178 (Lake Traverse). Underhill, 1959: 100 (introduced; throughout Vermillion R. basin).

The carp has been widely introduced and highly successful in South Dakota. It occurs in all major drainages and in most state waters except for the high-gradient streams of the Black Hills. Sooner or later it gains access to most reservoirs and stock ponds, often through introduction as bait.

Station records: 1, 4, 8, 21, 26B, 26C, 27, 33, 35, 38, 39, 43, 46, 49, 51, 53, 56, 58, 61, 64, 65, 72, 76, 83, 84, 127, 128, 135, 136, 137.

17. *Carassius auratus* (Linnaeus)—Goldfish

Carassius auratus. Churchill and Over, 1933: 37–38, fig. 23 (description, ecology, importance; Fall R., Hot Springs; Cody L., Todd Co.).

The goldfish is established in few South Dakota waters. Allum has seen examples only from Capitol Lake at Pierre and at Fall River, Hot Springs. It is reliably reported to be present in some stock dams, especially west of the Missouri River. Our survey collections yielded no specimens.

18. *Notemigonus crysoleucas* (Mitchill)—Golden shiner

Notemigonus chrysoleucas. Meek, 1892: 246 (Big Sioux R., Sioux City, Iowa [and S. D.]).
   Woolman, 1896: 357 (Wheatstone [Whetstone] Cr., Milbank).
Notemigonus crysoleucas. Churchill and Over, 1933: 43–44, fig. 28 (description, ecology, importance; east of the Missouri R.).

The old records by Meek and Woolman testify to the native occurrence of the golden shiner in eastern South Dakota. The species is scattered elsewhere, however, and records from west of the Missouri River are likely owing to artificial introductions.

Station records: 15, 33, 34, 37, 39, 40, 47, 84, 100.

COMMENTS ON THE GENERA *Semotilus* AND *Hybopsis*

Three species of North American minnows, *Cyprinus atromaculatus* Mitchill, *Cyprinus corporalis* Mitchill, and *Clinostomus margarita* Cope, share many characters, including two rows of hooked teeth with the count 5–4 in the main row, a complete lateral line, a terminal mouth of moderate size, scales of moderate size (about 43 to 78), a modal count of 8 anal rays, a common physiognomy, and, especially, a flaplike barbel located above the maxilla well in advance of its posterior tip. The intimate relationship of these species has been recognized by most recent workers (e.g., Jordan, Evermann, and Clark, 1930: 116), but usually each species has been assigned to a separate genus. The principal differences among the species relate to breeding behavior. Two construct nests, that of *corporalis* consisting of a large pile of pebbles and stones, that of *atromaculatus* an excavated depression in the gravel (Raney, 1949). The larger mouth in these species is employed to pick up and carry stones during nest construction, and the enlarged nuptial tubercles or pearl organs are used to drive off competitor males and egg predators and for protection while digging (Reighard, 1910: 1127–29; Raney, 1940). In *margarita*, which has a smaller mouth and has no enlarged tubercles, there is no movement of nest materials, and competing males that invade a territory are "promptly escorted away" (Langlois, 1929: 162), but apparently the male does not strike with the pearl organs as does *atromaculatus* (Reighard, 1910: 1128). It appears to be characteristic, though not invariable (Raney, 1940), for cyprinids that build and defend nests to develop enlarged nuptial tubercles and large mouths. The chubs of the subgenus *Nocomis* (genus *Hybopsis*) are similarly modified (Hankinson, 1932; Reighard, 1943), as are *Notropis cornutus* and the species of *Campostoma* (Raney, 1940). The strong selective advantage imparted by these structures appears to provide a rational explanation for their development. The presence of enlarged cephalic tubercles may be regarded as a specialization that has evolved independently several times among American cyprinids and catostomids. Their occurrence in some but not in all members of a kinship argues heavily against their overemphasis in classification. To stress such characters by granting generic status to monotypic or oligotypic phyletic lines, e.g., *Semotilus* for *atromaculatus*, *Leucosomus* for *corporalis*, *Margariscus* for *margarita*, and *Nocomis* for a small group of chubs that are so closely related that all were, until 1926, thought to belong to a single species (Hubbs, 1926: 27) is, we believe, to defeat the primary function of the genus in classification—the assembly of a group of related species within the most important supraspecific rank in the hierarchy of classification. We, therefore, agree with the recent treatment of *Nocomis* as a subgenus of *Hybopsis* (Bailey, 1951: 192; Hubbs and Lagler, 1958: 69) and of *Leucosomus* and *Margariscus* as synonyms of *Semotilus* (Hubbs and Lagler, 1958: 77–78). We see no useful purpose in the retention of *Semotilus*, *Leucosomus*, and *Margariscus* as subgenera; when the cyprinids are better understood the group here termed the genus *Semotilus* may well be ranked only as a subgenus.

KEY TO SPECIES OF *Semotilus*

1a.—No black spot on dorsal fin. Mouth smaller, upper jaw rarely extends to anterior margin of eye. Juveniles with lateral stripe on snout indistinct and diffuse, basicaudal spot typically nearly pinched off from lateral stripe. Breeding males with nuptial tubercles on head tiny, numerous; abdomen with a bright red or red-orange stripe. Lateral-line scales usually 63 to 78. Head length 3.4 to 4.1 (usually 3.5 or more) in standard length .... Pearl dace, *Semotilus margarita nachtriebi*
1b.—A black spot on dorsal fin at base of anterior rays (undeveloped in small young). Mouth larger, upper jaw extends beyond anterior margin of eye. Juveniles with lateral stripe on snout distinct, basicaudal spot broadly continuous with lateral stripe. Breeding males with nuptial tubercles on top of head greatly enlarged, few; abdomen without bright red stripe. Lateral-line scales usually 52 to 62. Head length 3.2 to 3.6 (usually 3.5 or less) in standard length........... Creek chub, *Semotilus atromaculatus*

19. *Semotilus margarita nachtriebi* (Cox) —Pearl dace
(Fig. 2)

Northern and western populations of the pearl dace have smaller scales than do those of the Allegheny region, as noted by Hubbs and Lagler (1958: 70). Counts of 38 specimens of *S. m. nachtriebi* range from 61 to 78, mean 69.0, and 95 per cent have 63 or more. Those of 37 specimens of *S. m. margarita* (Cope) from southern New York range from 49 to 63, mean 55.5, and 95 per cent have 62 or fewer.

S. m. nachtriebi, including the nominal *S. m. koelzi* (Hubbs and Lagler), is one of the most wide ranging of American cyprinids, occurring from the

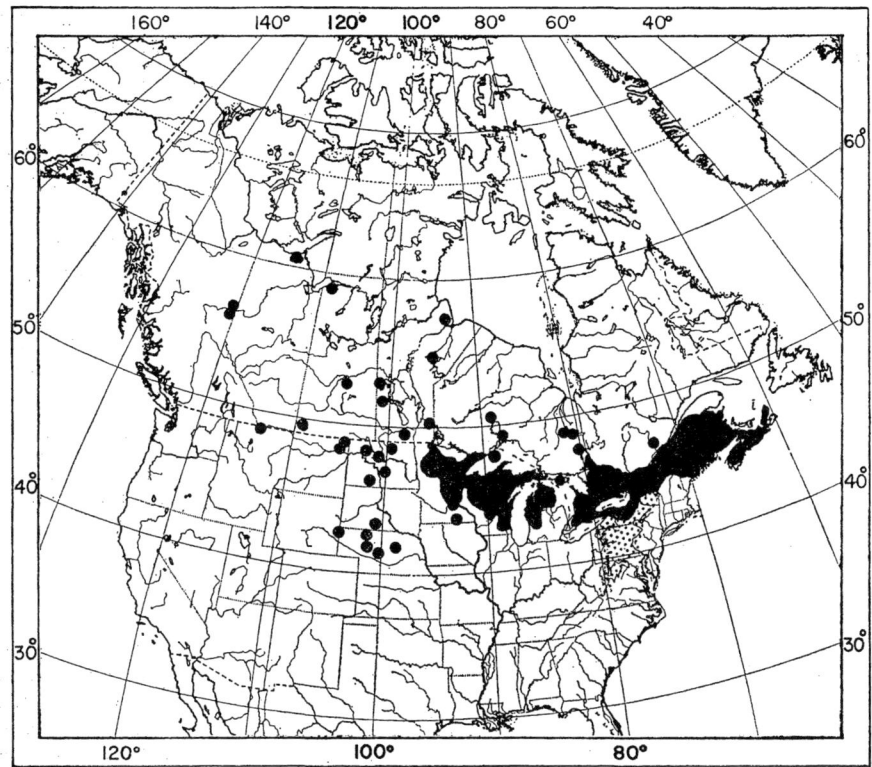

Fig. 2. Distribution of *Semotilus margarita*. Stippled area, *S. m. margarita*; solid spots and shaded area, *S. m. nachtriebi*. The paucity of records in much of Canada largely reflects collecting effort; the few spots in westcentral United States represent isolated relict populations.

Peace River drainage in British Columbia (Lindsey, 1956: 768) and the Athabaska River drainage across southern Canada, the northern United States, to Quebec, New Brunswick, Maine, and northern New York and Vermont (Fig. 2). In the northern Plains region the species is restricted to a few isolated relict populations, especially in cool, permanent spring waters of the sand hills of Nebraska and southern South Dakota. Here the species is a common associate of *Chrosomus neogaeus* and *C. eos*.

The Museum of Zoology has the following specimens from the Missouri River drainage:

Saskatchewan. UMMZ 164909, 69 mm. in st. l., trib. to Battle Cr., Milk R. dr., extreme southwest corner of province.

Montana. UMMZ 173923 (20 spec., 28 to 60 mm.), and UMMZ 173925 (2, 73–87 mm.), Wolf Cr., Roosevelt Co. UMMZ 173924 (7, 30–63 mm.), Ator Cr., Sheridan Co.

North Dakota. UMMZ 162340 (9, 66–102 mm.), Powers L., Burke Co. UMMZ 162346 (12, 51–97 mm.), Smedjik L. (drainage?), Burke Co. UMMZ 94765 (8, 64–85 mm.), trib. Curlew R., W Sims, Morton Co.

South Dakota. UMMZ 163812 (83 mm.), Rosebud L., Rosebud, Todd Co. [Sta. 93].

Nebraska. UMMZ 134508 (41, 32–74 mm.), Gordon Cr., N of Ray Cowles Ranch, 27 mi. N Hyannis, Cherry Co. UMMZ 134519 (8, 32–86 mm.), Snake R., 24 mi. S and 6 mi. E Merriman, Cherry Co. UMMZ 135754 (8, 27–35 mm.), UMMZ 135314 (50 mm.), UMMZ 87316 (29, 24–33 mm.), Beaver Cr., near head, 6 mi. N Bartlett, Wheeler Co. UMMZ 134217 (26 mm.), South Loup R., 1 mi. W Arnold, Custer Co.

20. *Semotilus atromaculatus* (Mitchill) —Creek chub

Semotilus macrocephalus. Girard, 1856: 40 (original description; [Missouri R.] Ft. Pierre, "Nebraska," UMMZ 56284*).

Leucosomus macrocephalus. Girard, 1858: 252–53 (description; [Missouri R.] Ft. Pierre, "Nebraska").

Semotilus corporalis (misidentification). Cope, 1879: 440 (Battle Cr. [? = Blue Blanket Cr.]).

Semotilus atromaculatus. Evermann, 1893b: 78 (Chicken and Crow crs. [Spearfish]; Rapid Cr. [Rapid City]). Evermann and Cox, 1896: 399–400 (description, habitat; Emanual Cr., Springfield; Crow Cr., [near] Chamberlain; Beaver Cr., Buffalo Gap; Rapid Cr., Rapid City; Redwater R., Crow and Chicken crs., Spearfish; Spring Cr., Hill City; French Cr., Custer; Belle Fourche R., Belle Fourche). Woolman, 1896: 348, 357 (Daugherty Cr., Browns Valley, Minn. [and S. D.]; Wheatstone [Whetstone] Cr., Milbank). Churchill and Over, 1933: 44–45, fig. 29 (description, ecology, importance; creeks of the state). Cleary, 1956: 88, 291 (description, life history, importance; Big Sioux R., Lyon [Lincoln], Sioux [Lincoln and Union], Plymouth [Union]), and Woodbury [Union] cos., Iowa [and S. D.]). Underhill, 1957: map 4 (Lake Traverse and Little Minnesota R. [Roberts Co.]; S Fork of Yellowbank R. [Grant Co.]). Hugghins, 1959: 21 (Medary Cr., College [= Six-mile] Cr.). Underhill, 1959: 100 (common in Vermillion R.).

One of the most common and widely distributed minnows in South Dakota, the creek chub has been taken by this survey or reported from all major drainages except the Little Missouri River. The species is usually an inhabitant of creeks, and no specimens have been taken from the Missouri River.

Station records: 3, 4, 5, 7, 8, 9, 14, 16, 18, 19, 24, 25B, 26B, 26C, 27, 28, 31, 34, 37, 39, 41, 43, 44, 48, 53, 59, 60, 61, 84, 86, 92, 93, 94, 104, 111, 116, 118, 127, 132.

KEY TO SPECIES OF *Chrosomus*

1a.—Teeth typically 2, 5–4, 2; the grinding surface small. Intestine short, with a single S-shaped loop. Only the midlateral dark stripe well defined. Basicaudal spot abruptly darker and narrower than axial stripe. Ventrolateral surface usually heavily peppered with melanophores. Angle of mouth extends almost to front of pupil..Finescale dace, *Chrosomus neogaeus*

1b.—Teeth typically 5–5; the grinding surface large. Intestine longer, with two or more loops or coils. Two lateral dark stripes well defined in adult (one only in young). Basicaudal spot scarcely or not narrower and only slightly darker than axial stripe. Ventrolateral surface immaculate. Angle of mouth does not reach eye............... .... Northern redbelly dace, *Chrosomus eos*

21. *Chrosomus neogaeus* (Cope)—Finescale dace
(Fig. 3)

Leuciscus neogaeus. Evermann, 1893b: 78 (Cox Lake and Chicken Cr. [Spearfish]). Evermann and Cox, 1896: 400 (description; Cox Lake, Spearfish).
  Pfrille neogaea. Churchill and Over, 1933: 54 (description, compiled; Cox Lake, NW Spearfish).

The finescale dace is widely distributed in the glaciated area of southern Canada and northern United States (Fig. 3). In the Missouri basin it occurs as a glacial relict, confined to cool spring waters where it is commonly associated with *Semotilus margarita nachtriebi* and *Chrosomus eos*. Hybridization between *eos* and *neogaeus* is common, and at some localities hybrids have been collected where one or both parent species were not taken. For example, the only evidence known to us that documents the occurrence of *Chrosomus neogaeus* in Colorado is a hybrid, *C. eos* $\times$ *C. neogaeus,* in the University of Kansas, 44.5 mm. in standard length, collected in Big Thompson River, at U. S. highway 34, 4.3 miles W of Loveland, Larimer County, by Olund, Metcalf, and Cross. The only indication of the presence of *C. neogaeus* in the Sand Hills area of southern South Dakota consists of four hybrids *C. eos* $\times$ *C. neogaeus*, UMMZ 127452, 38 to 55 mm. long (Sta. 85), that were taken with *C. eos*, but not *C. neogaeus*. In an ichthyological survey of Nebraska, Raymond E. Johnson took *C. neogaeus* at three localities in the Sand Hills, at all of which it was associated with *C. eos* or hybrids between these species, and another relict, *Semotilus margarita nachtriebi*. The data for the Nebraska collections of *Chrosomus neogaeus* are: UMMZ 87319 (26 mm.), and UMMZ 135753 (7, 27–30 mm.), Beaver Cr., 6 mi. N Bartlett, Wheeler Co. UMMZ 134504 (3, 35–38 mm.), Gordon Cr., N of Ray Cowles Ranch, 27 mi. N Hyannis, Cherry Co. UMMZ 134221 (2, 25–44 mm.), South Loup R., 1 mi. W Arnold, Custer Co.

George T. Baxter has sent for examination a series of *Chrosomus neogaeus* (University of Wyoming, Zoology Museum 2141, 9 specimens, 52–68 mm.) collected in Van Tassel Creek and the Niobrara River at the Nebraska line, Niobrara County, Wyoming. At this locality Dr. Baxter recorded *Semotilus margarita nachtriebi*, but he did not find *Chrosomus eos* or hybrids. We know of no records of *C. neogaeus* from Montana or North Dakota.

Cox Lake is really a gigantic spring hole lying north of the Black Hills

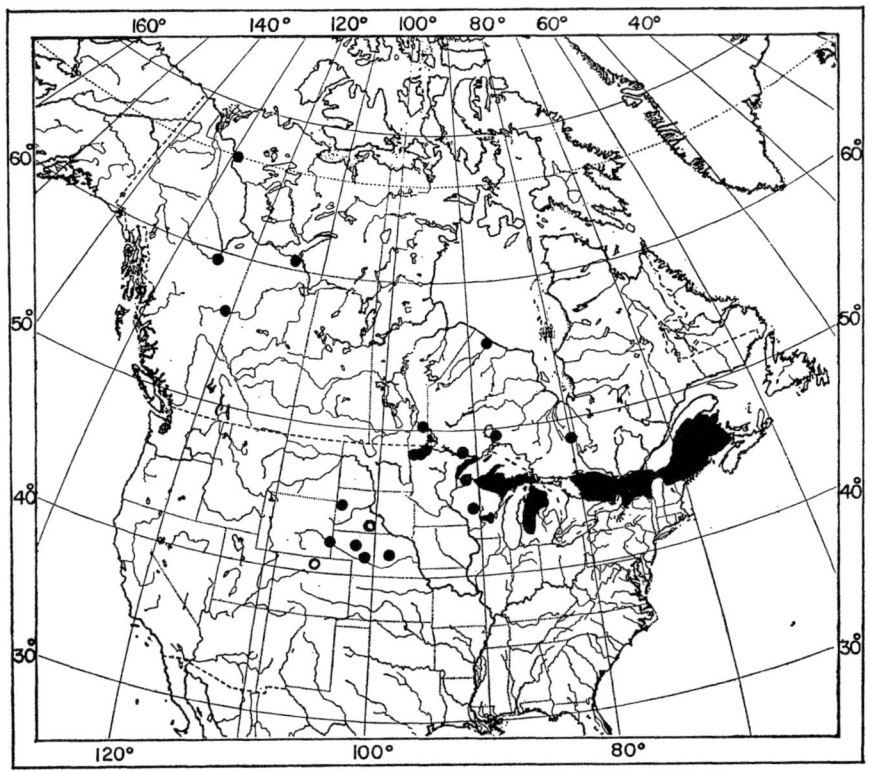

Fig. 3. Distribution of *Chrosomus neogaeus*. The open circles are localities for hybrids, *Chrosomus eos* × *C. neogaeus*, where *neogaeus* was not taken. The few records in most of Canada emphasize the need for more collecting; those in westcentral United States represent relict populations.

near Spearfish, South Dakota (station 124). Our examination reveals that it still supports the population of *Chrosomus neogaeus* reported by Evermann. The species abounds in Mud Lake, a small isolated spring pond nearby, and occurs in Mirror Lake and Redwater Creek, also in the immediate vicinity, and Evermann and Cox (1896b: 400) recorded its presence in nearby Montana Lake, near Beulah, Cook County, Wyoming. *Chrosomus eos* and *Semotilus margarita* have not been found at these localities.

Station records: 121, 122, 123, 124.

22. *Chrosomus eos* Cope—Northern redbelly dace

? *Chrosomus* sp. Cope, 1879: 440 (Battle Cr. [? = Blue Blanket Cr.]; presumptive identification).

Chrosomus dakotensis. Evermann and Cox, 1896: 395–96 (original description; Crow Cr., [near] Chamberlain). Churchill and Over, 1933: 53–54 (description, compiled; Crow Cr., near Chamberlain).

The northern redbelly dace is known from only three scattered areas in South Dakota—two adjacent spring-fed creeks in the Big Sioux drainage, a creek in the Niobrara drainage, and Crow Creek, an eastern tributary to the Missouri. Other relict populations occur in the Sand Hills of Nebraska: UMMZ 87318 (6, 20–28 mm.), Beaver Cr., 6 mi. N Bartlett, Wheeler Co. UMMZ 134503 (3, 35–42 mm.), Gordon Cr., N of Ray Cowles Ranch, 27 mi. N Hyannis, Cherry Co. Evermann and Cox (1896: 395–96) have reported the species (as *C. dakotensis*) from a pond at Niobrara [Knox Co.] and from Minnechaduza Creek at Valentine [Cherry Co.], Nebraska, and Underhill (1957: map 7) recorded it from southwestern Minnesota.

Station records: 25B, 28, 85.

KEY TO SPECIES OF *Hybopsis*

1a.—Teeth usually 2, 4–4, 2. Lateral-line scales 46 or more 2
  2a.—Head moderately compressed, deeper than broad; eyes lateral. Fins rounded; middle pectoral rays longest; pectoral not reaching insertion of pelvic. Scale rows above lateral lines in front of dorsal usually 25 to 27....... Lake chub, *Hybopsis plumbea*
  2b.—Head strongly depressed, broader than deep; eyes superolateral. Fins high and falcate; anterior pectoral rays longest; pectoral exceeding pelvic insertion in adult. Scale rows above lateral lines in front of dorsal usually 13 to 15..................... .... Flathead chub, *Hybopsis gracilis*
1b.—Pharyngeal teeth 0 or 1, 4–4, 0 or 1. Lateral-line scales 50 or fewer 3
  3a.—Mouth large, somewhat oblique, the premaxillae terminal or but slightly exceeded by snout, scarcely below lower border of eye. Breeding tubercles (in adult males), extending from between nostrils to occiput, large and sharp, directed forward. A red spot behind eye in adult. [Teeth 1, 4–4, 1.]................................ .... Hornyhead chub, *Hybopsis biguttata*
  3b.—Mouth smaller, horizontal, inferior, the premaxillae clearly exceeded by snout, and well below level of eye. Breeding tubercles covering most of head, minute, granular. No red spot behind eye. 4
    4a.—Teeth 1, 4–4, 1. Body not peppered throughout with macromelanophores 5
      5a.—Underside between pectoral and pelvic fins normally scaled. Eye large, contained 4 or less times in head. Gular area almost smooth, the sensory papillae minute. Adults 4 to 10 inches long. .... Silver chub, *Hybopsis storeriana*
      5b.—Ventral surface between pectoral and pelvic fins naked, or with scales only below pelvic bones. Eye small, contained 5 or more times in head. Gular area heavily papillose. Adults rarely more than 4 inches long 6
        6a.—Fins scarcely or not at all falcate; anterior dorsal rays usually exceeded by posterior rays in the depressed fin; pectoral fin not reaching insertion of pelvic (except in adult male). Body scales with prominent keels. Lateral-line scales 39 to 43. Belly naked. Head depressed; snout notably projecting, its length about equal to postorbital length of head...................... .... Sturgeon chub, *Hybopsis gelida*

6b.—Fins strongly falcate; anterior dorsal rays exceeding posterior rays in the depressed fin; pectoral fin reaching to or beyond insertion of pelvic. Scales without keels. Lateral-line scales 45 to 50. Belly with a few scales in prepelvic area. Head deeper and snout blunter, its length much less than postorbital length of head. .... Sicklefin chub, *Hybopsis meeki*

4b.—Teeth 4–4. Body with scattered macromelanophores. (*Hypothetical in South Dakota.*) .... Speckled chub, *Hybopsis aestivalis*

23. *Hybopsis plumbea* (Agassiz)—Lake chub

Couesius dissimilis. Evermann, 1893b: 78 (Rapid Cr. [Rapid City]*). Evermann and Cox, 1896: 410–12 (characters, variation, habitat; Crow Cr., [near] Chamberlain; Beaver Cr., Buffalo Gap; [Spring] Cr., Hill City; [French] Cr., Custer; Rapid Creek, Rapid City*). Churchill and Over, 1933: 55 (description, habitat, compiled; Beaver and Rapid crs.).

The lake chub is largely northern in distribution, but it occurs in the upper Missouri basin of Wyoming (Simon, 1946: 71; Personius and Eddy, 1955: 42) and Montana, and as a relict in North Dakota (UMMZ 162338, 103 mm., Powers Lake, Burke Co., Jan. 20, 1932, E. T. Judd), South Dakota, and the Niobrara drainage of Nebraska (Evermann and Cox, 1896: 410–12). The occurrence of the lake chub in the Sweetwater River of the Platte River drainage in Wyoming (Simon, 1946: 71) creates the suspicion of stream capture between the upper Bighorn or the Powder River and the North Platte River. An isolated population has been reported from the Mississippi drainage of northeastern Iowa (Bailey, 1956: 331).

In South Dakota the species is common in the streams of the Black Hills, occurs in the Little Missouri River, and has been reported from Crow Creek, an eastern tributary to the Missouri near Chamberlain.

Station records: 106, 109, 111, 113, 115, 136.

24. *Hybopsis gracilis* (Richardson)—Flathead chub

Pogonichthys communis. Girard, 1856: 24 (original description; [Missouri R.] Ft. Pierre, "Nebraska"). Girard, 1858: 247–48, pl. 55, figs. 1–6 (description; [Missouri R.] Ft. Pierre, "Nebraska").

Platygobio gracilis. Meek, 1892: 245 (description; Missouri R., Sioux City, Iowa [and S. D.]). Evermann, 1893b: 78 (Middle Cr. and Belle Fourche R. [Belle Fourche]; Hat Cr. [Ardmore]; Cottonwood Cr. [Edgemont]; "S. Fork of" Cheyenne R.* [Edgemont or Cheyenne Falls]). Evermann and Cox, 1896: 412 (habitat, coloration; White R., [near] Chamberlain; Redwater R., Spearfish; Cheyenne R.*, Edgemont and Cheyenne Falls). Churchill and Over, 1933: 45, fig. 30 [= *H. meeki*; apparently copied from Forbes and Richardson, 1909: fig. 45] (description, ecology, importance; White and Cheyenne rivers, tributaries, and Missouri R.).

Hybopsis gracilis. Cleary, 1956: 90, 292 (description, habitat, importance; Big Sioux R., Woodbury [Union] Co., Iowa [and S. D.]). Underhill, 1959: 100 (rare in Vermillion R. except at mouth).

Hybopsis gracilis gracilis. Olund and Cross, 1961: 330, pl. 24 (description, variation, synonymy; [Bad R. at Midland, Haakon Co.; White R., 6 mi. SW Belvidere, Washabaugh Co., and stations 64, 65, 69, 76, 88, 94, 120, 136 of this report]).

The flathead chub is the dominant minnow in the turbid, flowing waters of the Missouri River and the larger streams of the western half of South Dakota. It is absent in this area only from the higher streams of the Black Hills and from some small creeks. Ecologically it is largely replaced in slackwaters of rivers by *Hybognathus nuchalis* and *H. placitus.* East and north of the Missouri River the flathead chub occurs for only short distances upstream in larger tributaries.

A recent investigation of variation in *Hybopsis gracilis* (Olund and Cross, 1961) has revealed notable geographic differences in vertebral numbers. High counts were commonly encountered in the northern part of the range and in the Missouri-Mississippi River; low counts were characteristic of southern populations, in the Arkansas River and Rio Grande drainages, and those from some smaller streams on the Great Plains. The extreme types were assigned to different subspecies, *gracilis* and *gulonella,* respectively, and a large fraction of the range of the species was interpreted as consisting of intergrading populations. The morphological diversity was regarded as too great to be caused by the environment. But the vertebral counts are high in the north and in the larger, cooler rivers, precisely as would be predicted if low temperature during early development is effective in modifying vertebral numbers. In an attempt to clarify the picture we have made vertebral counts of 476 specimens of *Hybopsis gracilis* from South Dakota (Table 1). Our counts include four elements in the Weberian complex that were excluded by Olund and Cross. The shallow, exposed streams of western South Dakota yield consistently low counts. These waters characteristically have high temperatures during the summer breeding season of *H. gracilis,* and because of their shallowness diurnal fluctuations are presumably greater than in the Missouri River. The Missouri River, fed by mountain tributaries, remains relatively cool throughout the summer; late in the afternoon of August 24, 1952, the temperature at Station 64 was 72°F. immediately above the confluence with the Grand River, while a few feet away that of the Grand River was 79°F. During the breeding season of *H. gracilis* earlier in the summer when the Missouri River carries meltwater from mountain snowfields the differential is presumably greater. The mean vertebral number at this Missouri River station is 45.32; some 12 miles up the open watercourse of the Grand River (Station 135) the mean number is 43.36. Much farther upstream (Stations 132 and 131) the values are 42.63 and 42.79, respectively. On August 24, 1952, the Grand River at Station 135 carried a substantial flow and had occasional pools

TABLE 1

Frequency Distribution of Vertebral Counts of *Hybopsis gracilis* from South Dakota

Locality and Station	Number of Vertebrae							No.	Mean	S.D.	S.E.
	41	42	43	44	45	46	47				
Missouri R., Mobridge (64)	..	1	7	9	33	47	6	103	45.32	1.01	0.100
Missouri R., Pierre (65)	..	1	1	4	8	2	1	17	44.71	1.17	0.284
Missouri R., Ft. Randall (72)	..	..	..	5	12	12	1	30	45.30	0.79	0.144
Missouri R., Yankton (76)	..	1	1	2	15	25	5	49	45.57	0.96	0.137
Bad R., Midland (96)	3	13	10	4	..	..	..	30	42.50	0.87	0.159
Cheyenne R., Wasta (120)	..	29	27	3	1	..	..	60	42.60	0.66	0.086
Moreau R., N Dupree (129)	1	2	4	..	..	..	..	7	42.43	0.82	0.308
Moreau R., N Eagle Butte (130)	..	..	4	5	..	..	..	9	43.56	0.50	0.167
Grande R., Buffalo (131)	1	12	9	4	2	..	..	28	42.79	1.00	0.189
Grande R., Bison (132)	2	13	10	4	1	..	..	30	42.63	0.93	0.169
Grande R., NW Mobridge (135)	1	7	28	18	3	2	..	59	43.36	0.95	0.123
Box Elder Cr., Harding Co. (137)	1	16	28	9	..	..	..	54	42.83	0.71	0.097

up to eight feet deep. When seen here on July 27, 1955, it was dry for as far as could be seen from the bridge. It seems obvious that this stream must have been dry repeatedly during the severe drought of the middle 1930's, and it is equally certain that it was repopulated wholly or in large part from the Missouri River. The geographic pattern of variation is inconsistent with the view that vertebral numbers are under strict genetic control. We interpret the available data as strongly indicative of environmental modification of vertebral number in *Hybopsis gracilis*, but emphasize the desirability for thorough experimental testing of this hypothesis.

Station records: 64, 65, 66, 67, 69, 72, 73, 76, 79, 83, 87, 88, 91, 94, 96, 97, 98, 99, 102, 104, 117, 118, 120, 128, 129, 130, 131, 132, 135, 136, 137.

25. *Hybopsis biguttata* (Kirtland)—Hornyhead chub

Hybopsis kentuckiensis. Meek, 1892: 246 (Big Sioux R., Sioux Falls). Woolman, 1896: 348, 352, 357 (coloration; Daugherty Cr., and Little Minnesota R., Browns Valley, Minn. [and S. D.]; Wheatstone [Whetstone] Cr., Milbank).
Nocomis kentuckiensis. Churchill and Over, 1933: 46, fig. 31 (description, ecology, importance; tributaries of Big Sioux R.).
Hybopsis biguttata. Underhill, 1957: map 8 (Lake Traverse and Little Minnesota R. [Roberts Co.]; S Fork of Yellowbank R. [Grant Co.]).

Although our only collections of the hornyhead chub are from the

Minnesota River drainage it has been reported also from the Red and Big Sioux river systems.

Station records: 3, 4, 7, 8, 9.

26. *Hybopsis storeriana* (Kirtland)—Silver chub

The silver chub has apparently not been reported previously from South Dakota, but Cleary (1956: 292) has noted its occurrence in Rock River, a tributary to the Big Sioux River, in Sioux County, Iowa. In South Dakota it has been taken only in the Missouri River; the known upstream limit is in Gavins Point Reservoir (Lewis and Clark Lake).

Station record: 74.

27. *Hybopsis gelida* (Girard)—Sturgeon chub
(Fig. 4)

Hybopsis gelidus. Evermann and Cox, 1896: 409 (description, comparison with *H. meeki*; White R., [near] Chamberlain).

Macrhybopsis gelida. Churchill and Over, 1933: 55 (description, compiled; White R., near Chamberlain).

In its general distribution *Hybopsis gelida* lives in the larger streams of the northern Plains, where it is confined to the Missouri Drainage, and in the Missouri-Mississippi River proper at least as far downstream as Cairo, Illinois (Fig. 4). It ascends neither the Ohio River nor the Mississippi River above the confluence with the Missouri. Below Kansas City *H. gelida* is much less common than is *H. meeki*; in the middle part of the Missouri River both species are common, but in the upper part of the basin and in tributaries *H. gelida* occurs to the exclusion of *H. meeki*.

The sturgeon chub lives in a strong current, usually over a gravel bottom. Thus, though widely distributed in larger, heavily silt-laden streams of the Plains, it is spotty in occurrence since sand constitutes the predominant bottom material of the region. In South Dakota *Hybopsis gelida* is known only from the Missouri River and its larger western tributaries—the White, Cheyenne, and Grand rivers.

Figure 4 is based on the following distributional records:

Illinois. Mississippi R., bar 76, near Chester, Jackson Co. (UMMZ 164849). Mississippi R., Grand Tower, Jackson Co. (UMMZ 105473, 111576). Mississippi R., Cairo, Alexander Co. (UMMZ 147043).

Iowa. Localities listed by Harlan and Speaker (1956: 91).

Kansas. Localities reported by Cross (1953) and: Missouri R., T.7 S, R.21 E, sec. 12, Atchison Co. (KU). Missouri R., T.6 S, R.21 E, sec. 17, Atchison Co. (KU 3857). Kansas R., Eudora, Douglas Co. (KU 4488).

Missouri. Missouri R., Rocheport, Boone Co. (UMMZ 164819).

Fig. 4. Distribution by record stations of *Hybopsis gelida*.

Montana. Milk R. (Girard, 1856: 188; the type locality). Tongue R., Miles City (UMMZ 94171). Powder R., near mouth in Yellowstone R. (UMMZ 94170).

Nebraska. Bazile Cr. Niobrara (Evermann and Cox, 1896: 409). Missouri R., 3 mi. N and 2 mi. E Macy, Thurston Co. (UMMZ 135821). Missouri R., 3 mi. SE Plattsmouth, Cass Co. (UMMZ 134796). Platte R., 1 mi. S Gothenburg, Dawson Co. (UMMZ 92262, 106366). Platte R., Grand Island (Evermann and Cox, 1896: 409). Platte R. overflow pool, 3 mi. N and 3 mi. W Bellwood, Butler Co. (UMMZ 134682). Loup R., 1 mi. S Monroe, Platte Co. (UMMZ 135661). Elkhorn R., 1 mi. N Winslow, Dodge Co. (UMMZ 135772). Republican R., 1 mi. S Indianola, Red Willow Co. (UMMZ 135099). Republican R., 2 mi. E Inovale, Webster Co. (UMMZ 135055). Republican R., Superior, Nuckalls Co. (KU 2225).

North Dakota. Little Missouri R., Marmarth, Slope Co. (UMMZ 94759).

South Dakota. Station records: 64, 72, 76, 79, 88, 91, 120, 135; mouth of White R., near Chamberlain (Evermann and Cox, 1896: 409).

Wyoming. Big Horn R., 3 mi. below mouth of Shoshone R., Big Horn Co. (UMMZ 162937). Powder R., Arvada, and North Platte R., Douglas (Evermann and Cox, 1896: 409).

28. *Hybopsis meeki* Jordan and Evermann—Sicklefin chub
(Fig. 5)

Hybopsis gelidus (misidentification). Meek, 1892: 245 (description; Missouri R., Sioux City, Iowa [and S. D.]).
Hybopsis meeki. Bailey, 1951: 192 (Meek's specimens reidentified as *H. meeki*).

The sicklefin chub is known only from the Mississippi-Missouri River and the lower part of the Kansas River. It commonly occurs together with *Hybopsis gelida*. The known upstream limit is in the Missouri River at the mouth of the Little Missouri River; it ranges downstream to the mouth of the Missouri, thence in the Mississippi River at least to southeastern Missouri (Fig. 5). Like *H. gelida*, the sicklefin chub ascends neither the Ohio nor the Upper Mississippi River.

Hybopsis meeki lives in a strong current, typically over a sand bottom. It is admirably adapted with cutaneous sense organs for existence in a seemingly inhospitable environment (Moore, 1950). Its restriction to the excessively turbid waters of the Missouri-Mississippi raises a query as to its fate following impoundment of the Upper Missouri. If high turbidity is necessary to survival the species is probably already much reduced in the northern part of its range, but since turbidity remains high in the Missouri and Illinois part of this waterway, the species, which was formerly common there, presumably persists.

Figure 5 is based on the following distributional records:

Illinois. Mississippi R., near Grand Tower, Jackson Co. (UMMZ 105475, 111577, 147144). Mississippi R., Cairo, Alexander Co. (UMMZ 147042).

Iowa. Localities listed by Harlan and Speaker (1956: 91).

Kansas. Kansas R., below dam at Lawrence, Douglas Co. (KU 1842, 2172). Missouri R., T.7 S, R.21 E, sec. 12, Atchison Co. (KU 3848).

Missouri. Missouri R., St. Joseph (Jordan and Evermann, 1896: 317; type locality). Missouri R., 2.5 mi. SW Beverly Station on hwy. 92, Platte Co. (UMMZ 148878). Missouri R., 3.5 mi. N Courtney, by-pass hwy. U.S. 71, Jackson Co. (UMMZ 152574). Missouri R., 8 mi. E Lexington, Lafayette Co. (UMMZ 111700). Missouri R., 3 mi. NW Maltu Bend, Saline Co. (UMMZ 152608). Missouri R., Glasgow, Howard Co. (UMMZ 149024). Mississippi R., ¾ mi. below mouth Missouri R., St. Louis Co. (UMMZ 147091). Mississippi R., Cliff Cave, nr. Jefferson Barracks, St. Louis Co. (UMMZ 147131). Mississippi R., Crystal City, Jefferson Co. (UMMZ 147053). Plattin Cr., in pool connected to creek, 200 yds. from Mississippi R., Crystal City, Jefferson Co. (UMMZ 147049). Mississippi R., Brasher or Cottonwood Point, Pemiscot Co. (UMMZ 153076).

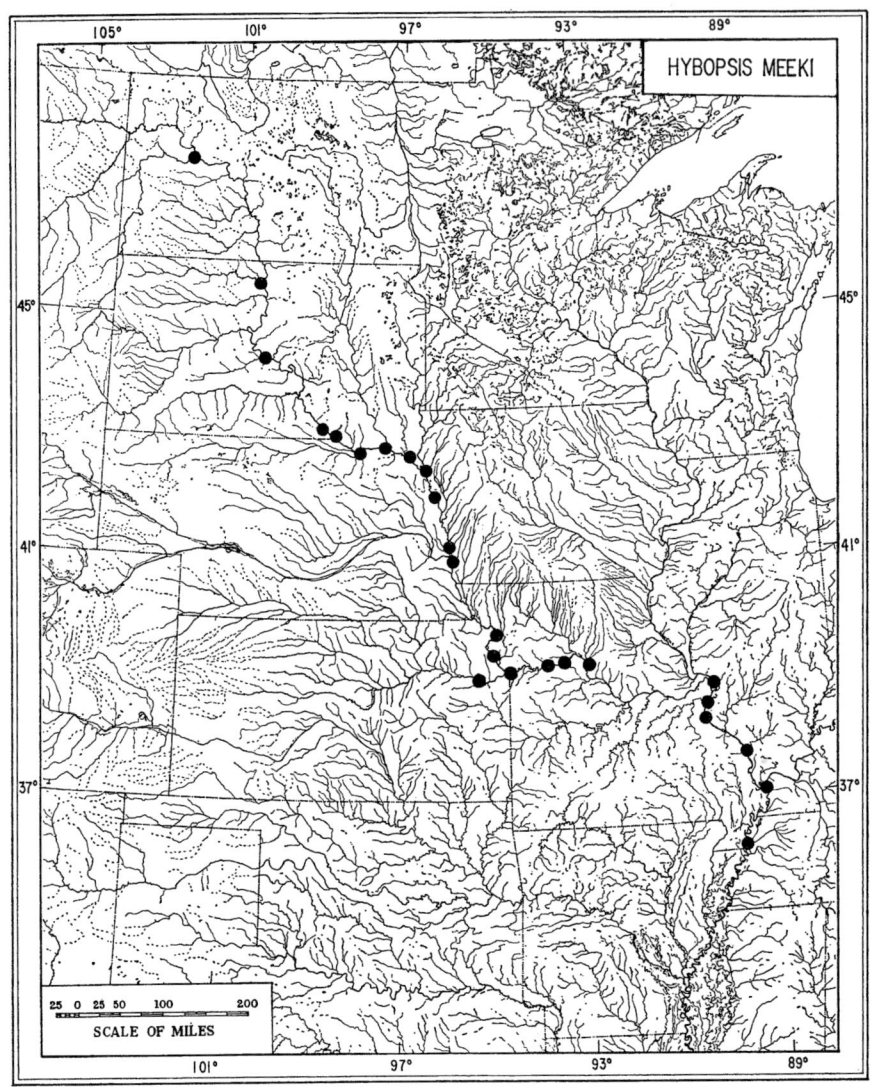

Fig. 5. Distribution by record stations of *Hybopsis meeki*.

Nebraska. Missouri R., 3 mi. N and 2 mi. E Macy, Thurston Co. (UMMZ 135822). Missouri R., 3 mi. SE Plattsmouth, Cass Co. (UMMZ 134795).

North Dakota. Missouri R. at mouth of Little Missouri R. (Personius and Eddy, 1955: 42).

South Dakota. Station records: 64, 65, 69, 72, 73, 76, 79.

KEY TO SPECIES OF *Rhinichthys*

1a.—Upper jaw scarcely exceeds lower jaw; mouth somewhat oblique. Lower lip as viewed from below of uniform width throughout. Dorsal origin equidistant from base of caudal and a point behind anterior margin of orbit, usually behind middle of pupil. Eye lateral, larger. Lateral dark stripe rather sharply defined below. Many dark scales scattered over body. .... Blacknose dace, *Rhinichthys atratulus meleagris*
1b.—Upper jaw greatly exceeds lower jaw; the nearly horizontal mouth sharklike in appearance. Lower lip as viewed from below widest posteriorly, tapered notably to its junction with mandibular ramus. Dorsal origin equidistant from base of caudal and a point anterior to rear edge of pupil, usually in front of orbit. Eye superolateral, smaller. Lateral dark stripe fades out gradually both above and below. Few or no scattered dark scales on body. .... Longnose dace, *Rhinichthys cataractae*

29. *Rhinichthys atratulus meleagris* Agassiz—Blacknose dace

Rhinichthys cataractae dulcis (misidentifications). Evermann and Cox, 1896: 408 (Crow Cr.*, [near] Chamberlain; Choteau Cr., Springfield).
Rhinichthys cataractae (misidentifications). Churchill and Over, 1933: 47 (description, ecology, importance; eastern streams of state; figure is of *R. cataractae*).
Rhinichthys atronasus. Churchill and Over, 1933: 47–48, fig. 33 (description, ecology, importance; eastern streams of the state).
Rhinichthys atratulus meleagris. Underhill, 1957: map 10 (Little Minnesota R. [Roberts Co.] and S Fork of Yellowbank R. [Grant Co.]). Hugghins, 1959: 22 (Medary Cr.). Underhill, 1959: 100 ([upper] Clay Cr. [Yankton Co.]).

Evermann and Cox (1896) failed to report *Rhinichthys atratulus meleagris* from South Dakota, but specimens from their collection have been examined by us from Enemy Creek, Mitchell, Davison County (UMMZ 86414) and from Crow Creek, near Chamberlain, Buffalo County (UMMZ 86397), and by Dr. Carl L. Hubbs from Choteau Creek, Springfield (MCZ). At Crow and Choteau creeks, but not at Enemy Creek, Evermann and Cox reported *Rhinichthys cataractae dulcis*. We have not confirmed the occurrence of the longnose dace in Crow and Choteau creeks or elsewhere in eastern South Dakota.

The blacknose dace occurs in the Minnesota, Big Sioux, Vermillion, and James river drainages as well as in Choteau and Crow creeks, all lying in eastern South Dakota. In Nebraska, Evermann and Cox (1896: 408) recorded the species, as *R. cataractae dulcis,* from a creek (the Verdigre River) at Verdigre and from Creighton Creek (Bazile Creek) at Niobrara (UMMZ 86395). In his survey of Nebraska fishes Dr. Raymond E. Johnson collected *R. atratulus meleagris* only in the East Branch of the Verdigre River, a tributary to the Niobrara River near its mouth, 6 mi. E and 6 mi. N of Royal, Antelope County (UMMZ 135321). These Nebraska localities are all close to the mouth of the Niobrara River. The only other localities of known occurrence in the Missouri basin are in southwestern Minnesota, (Underhill, 1957: map 10) and in northwestern Iowa (Cleary, 1956: 294; Larrabee, 1926: 14).

Station records: 3, 4, 5, 7, 8, 9, 18, 25B, 28, 29, 53.

30. *Rhinichthys cataractae* (Valenciennes)—Longnose dace

(?) *Rhinichthys maxillosus*. Cope, 1879: 440–41 (comparison; Battle Cr. [? = Blue Blanket Cr.]).

Rhinichthys dulcis. Evermann, 1893b: 78 (Whitewood Cr. [Deadwood]; Chicken and Crow crs. and Cooks Pond [Spearfish]; Cottonwood Cr. [Edgemont]; Hat Cr. [Ardmore]; Fall R. [Hot Springs]; Rapid Cr. [Rapid City]). Churchill and Over, 1933: 48 (description, ecology; western streams, Black Hills).

Rhinichthys cataractae dulcis. Evermann and Cox, 1896: 408–09 (variation, habitat; Cheyenne R., Cheyenne Falls* and Edgemont*; Cottonwood Cr., Edgemont; Chicken, Crow, and Spearfish crs. and Cox Lake, Spearfish; Beaver Cr., Buffalo Gap*; [Spring] Cr., Hill City; [French] Cr., Custer; Fall R., Hot Springs*; Whitewood Cr., Deadwood*; Rapid Cr., Rapid City). Woolman, 1896: 348 (Daugherty Cr., Browns Valley, Minn. [and S. D.]). [Cox (1897: 39) examined these specimens and compared them with some of Evermann's material from the Black Hills. The identification needs further checking.]

Rhinichthys cataractae. Hugghins, 1959: 22 (Rapid Cr.).

As indicated under the preceding species, Evermann and Cox's records of *Rhinichthys* from east of the Missouri River in South Dakota refer to *atratulus*. We know of no verified records of *cataractae* from there and therefore believe that Churchill and Over erred in citing eastern streams of the state in the range. We have not been able to check Woolman's (1896) report of *cataractae* from Daugherty Creek in the Lake Traverse basin, but suspect it is based on examples of *R. atratulus meleagris*, despite the occurrence of the longnose dace elsewhere in the Red River drainage (Underhill, 1957: map 11). West of the Missouri River in South Dakota the longnose dace has been taken in every major river system and is often common. We have not taken it in the Missouri River itself.

Some authors (e.g., Bailey, 1956: 357; Hubbs and Lagler, 1958: 68) have recognized that in *Rhinichthys cataractae* the swimbladder is rudimentary in the adult, and they have utilized this as a key character to assist in distinguishing *cataractae* from *atratulus*. Study of specimens of *cataractae* from the Black Hills reveals that, contrary to expectation, the swimbladder is typically well developed. Its total length is from 28.1 to 36.0 per cent of the standard length. The posterior tip extends to a point one-half to three-fourths the way back on the pelvic fin, or approximately to the posterior end of the body cavity. The posterior chamber of the swimbladder is large, subequal in diameter to the anterior chamber, and is not abruptly tapered posteriorly. Many additional specimens from scattered localities have been examined in an effort to elucidate the variation (Table 2). In the Black Hills, whether in streams or lakes, the swimbladder is uniformly well developed. In the Cheyenne River, into which the streams of the Black Hills drain, the swimbladder is consistently much reduced, is short and slender,

TABLE 2

DEVELOPMENT OF SWIMBLADDER IN *Rhinichthys cataractae*

Locality (Station or UMMZ Coll. No.)	S.L. (mm) Range	N	L. of Swimbladder (% of S.L.) Range	M	Posterior End of Swimbladder* (Frequency) $<P_2$	$=P_2$ to $\frac{1}{3}$	$>\frac{1}{3} P_2$	Post. Lobe (Ave. L.; % of S.L.)
Black Hills:								
Cox L. (Sta. 124)	53–70	10	30.1–36.0	32.0	0	0	10	20.9
Rapid Cr. (Sta. 115)	48–62	15	29.4–35.0	32.3	0	0	15	22.2
French Cr. (Sta. 106)	56–62	10	28.1–33.6	31.4	0	0	10	21.6
Cheyenne R.:								
Fall R., Hot Springs (177388)	50–62	7	23.1–34.8	29.3	0	3	4	18.8
At Oral (Sta. 104)	60–66	3	17.0–22.5	19.1	2	1	0	11.9
At Wasta (Sta. 120)	50–67	15	14.8–23.3	18.5	11	4	0	11.2
White R.:								
Little White R. (Sta. 94)	52–70	25	18.1–25.9	22.8	5	18	2	14.4
W Br. Rosebud Cr. (Sta. 92)	49–68	10	24.1–29.8	26.3	0	7	3	16.6
Little Missouri R.:								
Box Elder Cr. (Sta. 137)	52–60	5	27.4–37.0	30.6	0	1	4	20.5
Wyoming:								
Piney R., Sheridan Co. (161889)	48–73	14	13.9–24.3	20.1	5	8	1	12.9
Michigan:								
L. Superior, Luce Co. (68919)	53–73	15	20.0–26.9	23.7	1	13	1	15.2
L. Michigan, Big Stone Bay (88917)	54–98	15	24.0–32.2	27.9	0	8	7	18.0
Big Wolf Cr. (67889)	52–79	12	15.9–29.1	23.7	1	10	1	15.2
New Hampshire:								
Connecticut R., Coos Co. (126649)	52–87	8	19.5–29.8	23.9	4	3	1	14.6

* $<P_2$—Posterior end of swimbladder not reaching insertion of pelvic; $=P_2$ to $\frac{1}{3}$—extending to insertion of pelvic or to forward one-third; $>\frac{1}{3} P_2$—extending past forward one-third.

or the posterior chamber tapers posteriorly. Similar reduction is probably characteristic of adults through much of the range of the species as determined from examination of specimens from various localities. Measured samples from Little White River (sta. 94); Connecticut River, New Hampshire; Lake Superior, Michigan; and Big Wolf Creek, Alpena County, Michigan, differ but little from the Cheyenne River samples. Although there is slight overlap in length between extreme individuals and those from the Black Hills, the volume of the swimbladder falls into two discrete groups in these populations.

At some localities, however, the swimbladder varies widely within single populations, effectively bridging the gap between rudimentary and large bladders. In South Dakota an intermediate population occurs in the Little Missouri River; in the Upper Missouri drainage of Wyoming and Montana this seems to be a characteristic condition, and in a series from Big Stone Bay of Lake Michigan some specimens have well-developed swimbladders.

The young of *Rhinichthys cataractae* are midwater swimmers that customarily inhabit quiet pools or backwaters. The swimbladder is well formed. In streams, the most common habitat of larger fish, *cataractae* is primarily a riffle fish where it lives under stones in a strong or even torrential current. In this habitat the usual rudimentary swimbladder is of high adaptive value. Where it lives in large lakes the preferred environment is among rocks on a wave-swept shore, and here too a small swimbladder would appear to be desirable, but perhaps not essential to survival.

Why should fish from the Black Hills and Upper Missouri drainage have large or variably developed swimbladders? A possible explanation involves the absence of competition in the area. Creeks and small rivers of the region have few or no large predatory fishes and, perhaps more important, few other smaller carnivorous fishes. *Rhinichthys cataractae* has the capacity to survive in a varied and complex ecosystem by adaptation to a singularly inhospitable niche—the torrential riffle or wave-swept boulder shore. Here selection would quickly eliminate individuals with large swimbladders. But in a faunally impoverished area there would be no necessity to adopt such a restrictive habitat. If this supposition is correct, *Rhinichthys cataractae* should be much more versatile in its choice of habitat in the Upper Missouri basin than elsewhere in its range.

Station records: 84, 86, 87, 88, 92, 93, 94, 96, 97, 98, 102, 103, 104, 105, 106, 107, 108, 109, 112, 113, 114, 115, 116, 117, 120, 121, 123, 124, 127, 129, 130, 131, 132, 135, 136, 137.

31. *Phenacobius mirabilis* (Girard)—Suckermouth minnow

Phenacobius mirabilis. Evermann and Cox, 1896: 407–08 (characters; Crow Cr., [near] Chamberlain). Churchill and Over, 1933: 54–55, fig. 42 (compiled; Crow Cr. near Chamberlain).

The retention of the suckermouth minnow on the South Dakota faunal list depends solely on the record of Evermann and Cox from Crow Creek. In Nebraska Raymond E. Johnson did not find *Phenacobius mirabilis* in the Missouri River drainage above the mouth of the Platte River, but in Iowa, Cleary (1956: 295) recorded it from the Boyer River and from the Rock River, Sioux County, in the Big Sioux River drainage.

KEY TO SPECIES AND SUBSPECIES OF *Notropis*

1a.—Teeth in two rows, 1 or 2, 4–4, 1 or 2 2
  2a.—Principal anal rays 9 to 12 (occasionally 8 in *cornutus* and *illecebrosus*). Teeth 2, 4–4, 2. 3
    3a.—Origin of dorsal well behind insertion of pelvic, nearer base of caudal than tip of snout. Anal rays usually 10 to 12. 4
      4a.—Snout more blunt and shorter, its length usually contained more than 1.5 times in postorbital length of head. Eye larger, usually equal to or greater than snout. Body more compressed and deeper. Without rosy pigment............
.... Emerald shiner, *Notropis atherinoides*
      4b.—Snout sharp and produced, its length typically contained less than 1.5 times in postorbital length of head. Eye smaller, less than snout. Body thicker and more slender. Breeding males rosy about head and base of pectoral fin. ....
.... Rosyface shiner, *Notropis rubellus*
    3b.—Origin of dorsal ahead of to very slightly behind insertion of pelvic, nearer tip of snout than base of caudal. Anal rays usually 9 (often 8 in *illecebrosus*).... 5
      5a.—Dorsal fin very high, the anterior rays much exceeding posterior rays in the depressed fin and about equal to length of head. Exposed part of lateral scales not elevated, rounded behind. Predorsal scales about 15, not crowded or smaller than body scales. Interpelvic and prepelvic scales enlarged. Pelvic rays usually 9.
.... Silverband shiner, *Notropis illecebrosus*
      5b.—Dorsal fin of moderate height, the anterior rays not or but slightly exceeding posterior rays in the depressed fin, much shorter than head. Exposed part of lateral scales greatly elevated, diamond-shaped. Predorsal scales more than 20, crowded and much smaller than body scales. Interpelvic and prepelvic scales not enlarged. Pelvic rays usually 8. ...... Common shiner, *Notropis cornutus*
  2b.—Principal anal rays 7 or 8 (seldom 6 or 9; typically 9 and occasionally 10 in *lutrensis*, which never has 2, 4–4, 2 teeth). 6
    6a.—Body with a pronounced, black lateral stripe which passes through eye and surrounds snout; chin black. Lateral line incomplete..........................
.... Blackchin shiner, *Notropis heterodon*
    6b.—Body without a pronounced, black lateral stripe; chin unpigmented. Lateral line complete. 7
      7a.—Dorsal fin pointed in front, the anterior rays much exceeding posterior rays in the depressed fin. Eye larger, more than ¼ head length. Upper jaw straight

or gently curved (in lateral aspect). Scales usually not closely imbricated, the exposed surface not notably deeper than long. 8

8a.—Mouth moderately oblique, upper jaw forming an angle of more than 20° with the horizontal. Front of upper lip on level with bottom of pupil. Eyes lateral. Teeth 1 or 2, 4–4, 2 or 1 (usually with 2 teeth on one or both sides). 9

9a.—Anal rays typically 8. A large, circular, well-defined black spot at base of caudal fin. Dorsal fin higher, its depressed length contained 1.1 to 1.3 times in distance forward to occiput. ...Spottail shiner, Notropis hudsonius

9b.—Anal rays typically 7. No black spot at base of caudal fin. Dorsal fin lower, its depressed length contained 1.3 to 1.6 times in distance forward to occiput. .... River shiner, Notropis blennius

8b.—Mouth almost horizontal, upper jaw forming an angle of less than 15° with the horizontal. Front of upper lip on level with bottom of eye. Eyes superolateral. Teeth 1, 4–4, 1 (occasionally with tooth of minor row wanting on one side). .... Bigmouth shiner, Notropis dorsalis

7b.—Dorsal fin more or less rounded in front, the anterior rays much shorter than to slightly exceeding posterior rays (small juveniles) in the depressed fin. Eye smaller, less than ¼ head length in adult. Upper jaw with a definite (obtuse) angle near middle of its length. Scales more or less closely imbricated, exposed surfaces notably deeper than long. 10

10a.—Anal rays typically 8 (rarely 7 or 9). Scales usually 36 to 38. Body more elongate, its depth 3.6 to 4.1 in standard length. Dorsal (especially in adults) with a black blotch on membranes between posterior rays. Anal yellow in breeding males. Teeth usually 1, 4–4, 1. (*Hypothetical in South Dakota.*) .... Spotfin shiner, Notropis spilopterus

10b.—Anal rays usually 9 (often 8 or 10). Scales usually 34 or 35. Body deeper, its depth 2.7 (adults) to 3.7 (young) in standard length. Dorsal without black blotch. Anal red in breeding males. Teeth usually 4–4. .... Red shiner, Notropis lutrensis

1b.—Teeth in a single row, 4–4. 11

11a.—Anal rays usually 9 (often 8 or 10). Body depth 2.7 to 3.7 in standard length. Scales closely imbricate. .... Red shiner, Notropis lutrensis

11b.—Anal rays 7 or 8 (rarely 9). Body usually slender, depth 3.5 to 5.5 in standard length (3.0 or more in breeding individuals of N. topeka). Scales not closely imbricate, more or less rounded behind and loosely attached. 12

12a.—Anal rays typically 7 (rarely 6 or 8). 13

13a.—Mouth nearly horizontal, upper jaw forming an angle of less than 30° with the horizontal. Fins lower; length of depressed dorsal contained usually 2.2 to 2.3 times in predorsal length. Eye larger, contained less than 3.5 times in head length. Lateral stripe weakly developed, with at most an indistinct dark spot at base of caudal. Nuptial tubercles granular. Body and fins without red. .... Sand shiner, Notropis stramineus 14

14a.—Scale rows around body just in advance of dorsal and pelvic fins 21 to 27, usually 22 to 25. .... Notropis stramineus stramineus

14b.—Scale rows around body 24 to 37, usually 26 to 29. ....Notropis stramineus missuriensis

13b.—Mouth oblique, upper jaw forming an angle of over 30° with the horizontal. Fins higher; length of depressed dorsal usually 1.8 to 2.0 times in predorsal length. Eye smaller, contained more than 3.5 times in head. A prominent, lateral dusky stripe terminating at base of caudal in a distinct, though small, dark spot. Nuptial tubercles on head coarse and sharp. Nuptial males with the fins and lower side bright red or orange. .... Topeka shiner, *Notropis topeka*

12b.—Anal rays typically 8 (rarely 7 or 9). 15

15a.—Mouth small, strongly oblique, upper jaw forming an angle of more than 45° with the horizontal. Chin dark; peritoneum dark. Anterior lateral-line scales rounded behind, not elevated. Infraorbital canal complete. *(Hypothetical in South Dakota.)* .... Pugnose shiner, *Notropis anogenus*

15b.—Mouth moderate, little oblique, upper jaw forming an angle of less than 30° with the horizontal. Chin pale; peritoneum silvery. Anterior lateral-line scales visibly elevated. Infraorbital canal interrupted, usually in three sections. .... Blacknose shiner, *Notropis heterolepis*

32. *Notropis atherinoides* Rafinesque—Emerald shiner

Notropis dilectus. Meek, 1892: 245–46 (description; Big Sioux R., Sioux Falls, and Missouri R., Sioux City, Iowa [and S. D.]). Evermann and Cox, 1896: 407 (description, habitat; James R.*, Mitchell). Churchill and Over, 1933: 54 (compiled; James R., Mitchell).

Notropis atherinoides. Meek, 1892: 246 (Big Sioux R., Sioux City, Iowa [and S. D.]). Woolman, 1896: 351 (coloration; Big Stone L., Creagers Farm, Minn. [and S. D.]). Cleary, 1956: 296 (Big Sioux R., Lyon* [Lincoln] and Sioux [Lincoln and Union] cos., Iowa [and S. D.]). Underhill, 1959: 100 (lower Vermillion R. and mouth).

Notropis percobromus. Hubbs, 1945: 17 (distribution, characters; Missouri R. system, South Dakota). Hubbs and Bonham, 1951: 93–95 (comparisons; distribution).

Notropis rubellus (misidentification). Cleary, 1956: 296 (Big Sioux R.*, Lyon [Lincoln] Co., Iowa [and S. D.]).

Alburnellus percobromus Cope, 1871, described from St. Joseph, Missouri, has a troubled nomenclatural history which Hubbs (1945) and Hubbs and Bonham (1951: 93) have brought to its current status by recognition as a full species, *Notropis percobromus*, related to *Notropis atherinoides*. *N. percobromus* was said to range from the Red and Arkansas river systems in the Great Plains of Oklahoma and Kansas to the Mississippi River in Tennessee, northward through Nebraska, Missouri and Iowa to the Missouri River system in the Dakotas and to the Mississippi River between Minnesota and Wisconsin. This judgment has been followed by most workers (e.g., Eddy and Surber, 1947; Eddy, 1957; Moore, 1957), but Bailey (1951: 192) questioned whether *percobromus* was merely a subspecies of *atherinoides* and later (1956: 332) expressed doubt as to the genetic distinctness of these nominal species. At that time he reidentified Minnesota specimens supposed to be *percobromus* as *atherinoides*. Subsequently Hubbs and Lagler (1958: 81) have reaffirmed their confidence in the specific status of these forms which " . . . are readily distinguishable where they occur together." No data were presented to support this contention, which apparently rests on the comparison published by Hubbs (1945: 17). That comparison, though not so stated, was drawn from collections taken in the Poteau River at Slate Ford, near Shady Point, Oklahoma. On December 1, 1939, John D. Mizelle took 149 specimens, 32 to 55 mm. in standard length, identified as *Notropis atherinoides* (UMMZ 127302), and 48 specimens, 31 to 43 mm., identified as *N. percobromus* (UMMZ 127294). A prior collection, taken from the same locality by George A. Moore, August 8, 1939, also contains *N. atherinoides* (UMMZ 127207, 206 spec. 24–49 mm.) and *N. percobromus* (UMMZ 127210, 4 spec. 30–41 mm.). A subsequent collection by Moore was taken on April 12, 1941, after the comparison was drafted. It also contains *N. atherinoides* (UMMZ 137897, 382 spec. 28–62 mm.) and *N. percobromus* (UMMZ 137895, 4 spec., 19–27 mm.). Carl L. Hubbs participated in the identification of all of these fish.

We have re-examined these collections and agree with Dr. Hubbs that most specimens of the *atherinoides* complex represented therein can be separated into two groups. Several of the measurable characters mentioned in Hubbs' (1945) comparison are not reliable, or are subject to very broad overlap (Table 3); but pigmentation, especially the darker lips of *"atherinoides,"* provides a satisfactory means for separating the groups, and we assume that Hubbs used this in his sorting. Some other characters show group consistency even though they do not permit reliable identification of individuals. For example, mean vertebral counts of the two groups (Table 4) differ by 0.57 vertebra, a figure that is significant (t value = 2.549 with 49 d.f.; $p = <.02$).

In disagreement with Dr. Hubbs, we do not believe these groups represent separate species. For years we have tried without success to apply the characters in his comparison to identify fish of the *atherinoides* complex. Pigmentation varies greatly with locality and is useless. Plains collections average somewhat chubbier in body form, as measured by head depth, body depth, or peduncle depth, and the head is usually a bit longer (Table 3). But these and other characters vary with locality and are not geographically consistent. Head length decreases proportionally with age, and the largest specimen of *"percobromus"* in the Poteau series is only 43 mm. long. We have taken numerous measurements and plotted scattergrams in a vain effort to discover satisfactory distinguishing characters. These data are summarized in Table 3 and are on file in the Museum of Zoology. It is evident that the measurable characters suggested by Hubbs fail to provide an adequate distinction between species. How then is one to explain the occurrence in the Poteau River (and in at least one other locality where a similar situation has been encountered) of two apparently distinct groups of individuals?

Intensity of pigmentation in fishes is known to vary over wide limits

TABLE 3

SUMMARY OF MEASUREMENTS OF *Notropis atherinoides* CONTRASTED WITH HUBBS' (1945) COMPARISON OF *N. atherinoides* AND *N. percobromus* PROPORTIONS ARE EXPRESSED AS THOUSANDTHS OF STANDARD LENGTH

Locality and UMMZ Number	N	Size Range (mm)	S.L.	Head L.	Orbit L.	$\frac{\text{Head L.}}{\text{Head D.}}$	Means $\frac{\text{Dorsal to occiput}}{\text{Dorsal height}}$	Head D.	Greatest D.	Least D.
Okla.: Poteau R. *N. atherinoides*: Hubbs				<250	Larger	ca. 1.6	1.8–2.1			
Lake Erie (159808)	10	36–39	38.0	251	086	1.60	1.78	157	196	089
	10	40–44	42.7	240	084	1.57	1.87	153	194	086
	10	45–65	54.4	238	082	1.58	1.86	151	194	094
	30	36–65	45.0	243	084	1.58	1.84	154	194	090
S. D.: Whetstone Cr., (167001)	10	63–81	68.8	230	072	1.62	1.91	141	201	094
Okla.: Poteau R., (127302) "*atherinoides*"	26	35–45	40.3	244	087	1.64	1.79	149	190	091
	24	46–55	48.7	236	084	1.65	1.89	144	188	094
	50	35–55	44.3	240	086	1.64	1.84	146	189	092
Okla.: Poteau R. *N. percobromus*: Hubbs				>250	Smaller	ca. 1.4	1.4–1.85			
Mont.: Missouri R. dr. (173916, 173920)	24	28–73	60.3	248	077	1.58	1.84	153	209	096
N. D.: Missouri R. dr. (94777, 94779)	14	32–82	39.7	252	083	1.56	1.65	162	217	093
Nebr.: Platte R. dr. (134705, 134732, 135657, 135717)	33	33–63	43.4	258	086	1.57	1.75	164	224	104
Kan.: Arkansas R. dr. (159768, 160435)	13	41–66	54.6	262	078	1.55	1.89	169	234	103
Okla.: Poteau R. (127294) "*percobromus*"	26	36–43	38.4	259	088	1.59	1.69	164	215	102

depending on environmental factors such as turbidity, and since the most obvious and consistent differences between these two groups involve pigmentation we suspected that the Poteau River might have two distinct and unlike habitats. Dr. George A. Moore informs us that such is the case; adjacent to the Poteau River there are oxbow lakes and overflow pools. One pool at Slate Ford, when Moore last saw it, was about waist deep and 100 feet long by 30 or 40 feet wide; it was quite productive of fishes. All of the so-called *percobromus* in these collections are small and pale. We judge that these fish came from the more turbid of the habitats. The different environmental conditions in pond and stream, not only instantaneously but throughout growth and development of the fish, offer a rational explanation for the partial separation of the two types in proportions (Martin, 1949) and meristics. It is noted that the average for "*percobromus*" is 38.16 vertebrae, for "*atherinoides*" 38.73. Vertebral counts from elsewhere in the range of *atherinoides* (Table 4) yield a clinal gradient with higher numbers in the north, whether in the Missouri basin ("*percobromus*"), the Mississippi basin, or the Great Lakes ("*atherinoides*"). Mr. Glenn Flittner, who is studying variation of *N. atherinoides* in the Great Lakes, finds that at Isle Royale, Lake Superior, near the northern limit of thermal tolerance of the species, the mean vertebral count is 41.05, the highest known for the species. If a species has a brief spawning period correlated with a rather

TABLE 4

Frequency Distribution of Vertebral Counts of *Notropis atherinoides*

Locality and UMMZ No.	Number of Vertebrae						N	M	S.D.	S.E.
	37	38	39	40	41	42				
Ohio: Lake Erie (159808)			3	16	9	1	29	40.3	0.707	0.131
S. D.: Big Stone L. drainage (167001, 167132)				8	6	1	15	40.5	0.655	0.169
Mont.: Marias R. (173920)			2	5	3	..	10	40.1	0.745	0.236
Missouri R. (173916)			2	3	..	..	5	39.6	0.500	0.224
N. D.: Cedar R. (94779)			9	1	..	..	10	39.1	0.333	0.105
S. D.: James R. (127416)	2	7	1	1	..	..	11	38.1	0.837	0.252
Nebr.: Clear Cr. (134705, 134732)	1	13	6	..	..	..	20	38.2	0.562	0.126
Kans.: Arkansas R. (160435)	1	6	..	..	..	..	7	37.9	0.408	0.154
Spring Cr. (159768)	4	2	..	..	..	..	6	37.3	0.447	0.183
Okla.: Arkansas R. (113365)	3	11	6	..	..	..	20	38.2	0.688	0.154
Poteau R. "*percobromus*" (127294)	7	8	9	1	..	..	25	38.2	0.890	0.178
Poteau R. "*atherinoides*" (127302)	..	10	13	3	..	..	26	38.7	0.663	0.130

definite temperature, little environmental modification of vertebral count may be anticipated. But Mr. Flittner finds that *atherinoides* has a protracted summer breeding season with wide thermal limits, factors conducive to marked environmental control.

We conclude that the differences in two stocks of the *atherinoides* group taken at the Poteau River are attributable to environmental factors. Since evidence for the specific identity of *percobromus* rests on the supposed sympatric occurrence of two species here, we place *Alburnellus percobromus* Cope in the synonymy of *Notropis atherinoides*. There are clinal gradients in vertebral number and perhaps weak clines in some body proportions in *atherinoides* that probably reflect direct environmental control, but there are no evident grounds for the recognition of subspecies.

The status of *Notropis oxyrhynchus* and *N. jemezanus* needs to be examined carefully since they are allopatric representatives of the *atherinoides* group that may prove to be conspecific with it. *N. oxyrhynchus* from the Brazos River was compared with *percobromus* and *atherinoides* (Hubbs and Bonham, 1951); the snout in *oxyrhynchus* averages sharper, but in this character it closely approaches *N. jemezanus* of the Rio Grande basin, a nominal species that was not compared. If not the same, these three forms are intimately related.

Notropis atherinoides ranges from the Gulf Coastal Plain of the Trinity River in eastern Texas to the Alabama River basin, north, west of the Appalachians, through the Mississippi basin and St. Lawrence drainage to Lake Champlain and Quebec, west through southern Canada to Great Slave Lake, Northwest Territories, and south through Alberta, Montana, and the Dakotas to the Red River drainage of northern Texas.

In South Dakota the emerald shiner occurs in the Minnesota, Big Sioux, Vermillion, and James river drainages. The species was probably uncommon originally in the Missouri River, but following impoundment it has increased and will likely become a dominant species in the reservoirs. We have examined specimens collected recently in Fort Randall Reservoir.

Station records: 3, 5, 8, 15, 16, 22, 26A, 26B, 51.

33. *Notropis rubellus* (Agassiz)—Rosyface shiner

Notropis rubellus. Underhill, 1957: 16, map 14 (Little Minnesota R. [Roberts Co.] and Big Stone L. [Grant Co.]).

The rosyface shiner is known in South Dakota only from the Minnesota River drainage, but may occur also in the Red River system. Cleary (1956: 296) mapped *rubellus* as occurring in the Big Sioux River at the extreme northwest corner of Iowa, his only record from the Missouri drain-

age. This record is probably based on a specimen of *N. atherinoides* (UMMZ 167970) sent to the Museum of Zoology by Cleary for identification shortly before his report appeared. He did not map *atherinoides* at that locality.
Station record: 7.

34. *Notropis illecebrosus* (Girard)—Silverband shiner

(?) *Notropis dilectus* (misidentification). Meek, 1892: 245 (in part; Missouri R., Sioux City, Iowa [and S. D.]).
Notropis illecebrosus. Bailey, 1951: 192 (Missouri R., Sioux City, Iowa). Bailey, 1956: 332 (habitat; accuracy of Sioux City record questioned). Underhill, 1959: 100 (mouth of Vermillion R.).

The recent discovery of the silverband shiner at the mouth of the Vermillion River and the additional specimens here reported from the Missouri River confirm the questioned occurrence of the species at Sioux City. *Notropis illecebrosus* presumably lives in suitable habitats in or near the Missouri River to its mouth, but there are few records. The Museum of Zoology has only the following from this basin: *Kansas*: Missouri R., Leavenworth (No. 87099). *Missouri*: Petite Saline Cr. into Missouri R., 3.5 mi. SE Boonville, Cooper Co. (150581); Tuque Cr., 1 mi. SE Marthasville, Warren Co. (147928); Femme Osage Cr., 2 mi. NE Defiance, St. Charles Co. (147900). Recent collections from Oahe Reservoir, in the Spring Creek area (T. 112 N, R. 80 W, sec. 6), Hughes Co. (UMMZ 178860), and along the east shore 18 mi. N of Oahe Dam, Sully Co. (UMMZ 178867), both in South Dakota, constitute the upstream and northernmost known occurrences.
Station records: 68, 75, 80, 81, 82.

35. *Notropis cornutus* (Mitchill)—Common shiner

Notropis megalops. Meek, 1892: 246 (Big Sioux R.*, [Sioux City, Iowa or Sioux Falls, S. D.]). Woolman, 1896: 348, 351, 357 (coloration, comparisons; Daugherty Cr., Browns Valley, Minn. [and S. D.]; Little Minnesota R.*, [Sisseton] Indian Agency, S. D., and Browns Valley, Minn. [and S. D.]; Big Stone L., Creagers farm and Ortonville, Minn. [and S. D.]; Wheatstone [Whetstone] Cr., Milbank).
Notropis cornutus. Evermann and Cox, 1896: 405–06 (description; Rock, Enemy, and Firesteel crs., Mitchell; Prairie Cr.*, Scotland; Choteau and Emanuel crs., Springfield). Hugghins, 1959: 22 (Medary Cr., Big Sioux R.). Gilbert, 1961: 189 (distribution records mapped).
Luxilus cornutus. Churchill and Over, 1933: 52–53, fig. 41 (description, ecology, importance; streams of eastern part of state, rarely in lakes).
Notropis cornutus frontalis. Cleary, 1956: 297 (Big Sioux R., Lyon [Lincoln] and Sioux [Lincoln and Union] cos., Iowa [and S .D.]). Underhill, 1957: map 15 (Little Minnesota R. and Lake Traverse [Roberts Co.]; S Fork of Yellowbank R. [Grant Co.]. Underhill, 1959: 100 (Vermillion R.).

Notropis cornutus occurs in all principal drainage areas east of the Missouri River, but we know of no records of the presence of the species in or west of that river in South Dakota or anywhere in the Missouri basin above Choteau Creek. Gilbert (1961) has recently synonymized the subspecies *frontalis* with the typical subspecies.

Station records: 3, 4, 7, 8, 9, 11, 16, 18, 26A, 26B, 26C, 27, 28, 29, 31, 39.

36. *Notropis heterodon* (Cope)—Blackchin shiner

Notropis heterodon. Woolman, 1896: 357 (Wheatstone [Whetstone] Cr., Milbank).
Hybopsis heterodon. Churchill and Over, 1933: 49, fig. 35 (lakes and streams east of Missouri R.).

The above report of the blackchin shiner from Whetstone Creek in the Minnesota River drainage is in line with the known distribution of the species, but the general statement by Churchill and Over cannot be accepted without confirmation. The only verified occurrence of *Notropis heterodon* in the Missouri River basin known to us is based on a specimen from Spirit Lake, Dickinson County, Iowa, collected in 1932 (UMMZ 101537; Harlan and Speaker, 1956: 96). Meek's report (1892: 247) from nearby Silver Lake probably is correct, but may rest on specimens of *Notropis anogenus*. We have not seen specimens from South Dakota.

37. *Notropis hudsonius hudsonius* (Clinton)—Spottail shiner

Notropis hudsonius. Meek, 1892: 246 (Big Sioux R., Sioux City, Iowa [and S. D.]). Evermann and Cox, 1896: 404 (characters; James R.*, Rock and Firesteel crs., Mitchell). Woolman, 1896: 351–52 (description; Little Minnesota R., Browns Valley, Minn. [and S. D.]; Big Stone L., Creagers farm, Minn. [and S. D.]).
Hudsonius hudsonius. Churchill and Over, 1933: 51, fig. 38 (description, ecology, importance; lakes and larger streams east of Missouri R.).

The spottail shiner occurs in the Missouri River system only in the James drainage, the Big Sioux drainage of South Dakota and Minnesota (Underhill, 1957: map 15), and in lakes and streams of northwestern Iowa (Cleary, 1956: 298). It occurs also in the Minnesota River drainage, from which it presumably made its way into the Missouri drainage in postglacial time.

Station records: 3, 5, 16, 22, 24, 51.

38. *Notropis blennius* (Girard)—River shiner

Paranotropis jejunus. Churchill and Over, 1933: 51–52, fig. 39 (description, ecology, importance; streams east of Missouri R., occasionally in lakes).
Notropis blennius. Underhill, 1959: 100 (mouth of Vermillion R.).

Despite the comment by Churchill and Over, the river shiner is probably confined in South Dakota to the Missouri River and the lower parts of its larger tributaries. We have not collected or examined specimens. Raymond E. Johnson's survey of Nebraska fishes yielded no records from the Missouri River or tributaries above the mouth of the Platte River.

39. *Notropis dorsalis* (Agassiz)—Bigmouth shiner

Notropis dorsalis. Cleary, 1956: 299 (Big Sioux R., Lyon [Lincoln] and Sioux [Lincoln and Union] cos., Iowa [and S. D.]). Underhill, 1957: map 18 (Little Minnesota R. and Lake Traverse [Roberts Co.]; S Fork of Yellowbank R. [Grant Co.]). Moyle and Clothier, 1959: 179 (Lake Traverse). Underhill, 1959: 100 (Vermillion R. and headwaters of Clay Cr.).

The bigmouth shiner is widely distributed in eastern South Dakota, where taken in all major drainages; it occurs also in Antelope Creek of the Niobrara drainage and was caught once in the Missouri River. The characteristic habitat is small streams with a sandy bottom. The failure of early collectors to note this now common species suggests that the bigmouth shiner has increased in abundance and has extended its range during recent decades.

Station records: 3, 5, 7, 8, 9, 11, 16, 18, 19, 24, 25B, 26C, 27, 28, 29, 45, 49, 53, 73, 84.

40. *Notropis lutrensis* (Baird and Girard) —Red shiner

Notropis whipplei (misidentification). Meek, 1892: 246 (Big Sioux R., Sioux City, Iowa [and S. D.]).
Notropis lutrensis. Evermann and Cox, 1896: 404–05 (description, distribution; James R., Enemy and Rock crs., Mitchell; Crow Cr.*, [near] Chamberlain; Emanuel and Choteau crs., Springfield). Cleary, 1956: 300 (Big Sioux R., Lyon [Lincoln] and Sioux [Lincoln and Union] cos., Iowa [and S. D.]). Underhill, 1959: 100 (Vermillion R., from Parker to mouth).
Moniana lutrensis. Churchill and Over, 1933: 52, fig. 40 (description, ecology, importance; streams east of Missouri R., occasionally in lakes).

The red shiner is known in South Dakota only from the Missouri River (where rare) and its tributaries from the east and north (where common). Meek's record of *Notropis whipplei* at Sioux City was based on *N. lutrensis*, as confirmed by examination of the material in the Museum of Comparative Zoology by Carl L. Hubbs.

Station records: 24, 26A, 26B, 26C, 27, 29, 32, 34, 35, 43, 44, 51, 53, 55, 56, 61, 66, 72, 73, 83.

41. *Notropis stramineus* (Cope)—Sand shiner

The sand shiner, long incorrectly known as *Notropis blennius* (Hubbs, 1926: 43), has recently been called *Notropis deliciosus*, a name that now must be placed in the synonymy of *Notropis texanus* (Suttkus, 1958). The sand shiner becomes *Notropis stramineus*.

Populations of this species from South Dakota fall into two sharply defined morphotypes characterized by high (mean values 27.1 to 31.0) and low (23.6 to 25.0) counts of body-circumference scales (Table 5). All collections from the Minnesota, Big Sioux, Vermillion, and James river drainages, in the eastern part of the state, have low counts; all others, including lesser tributaries of the Missouri River (Emanuel and Choteau creeks) near the mouth of the Niobrara River a short distance above the James River, have high counts. Within either region, counts vary from station to station without marked geographic correlation. For example, two samples from the Little Missouri drainage average 27.1 and 28.8 scales. The counts for western South Dakota are in close agreement with those for the Upper Missouri drainage of Montana, North Dakota, and northern Nebraska at least as far east as the Bow River, Cedar County, just downstream from the mouth of the James River. Notably high counts are characteristic of the upper Niobrara system. The contrasting lower counts of the rivers of eastern South Dakota are matched by those in the Little Sioux drainage of Iowa and the Red River system in North Dakota. In the region under consideration there appears to be little gene flow between the two types. It seems likely that the smaller-scaled western type, *Notropis stramineus missuriensis*, has repopulated most of the Missouri basin from the Plains region to the south, and that the larger-scaled *N. s. stramineus* has crossed into that part of the Missouri basin between the James and the Little Sioux rivers from the upper Mississippi basin, presumably from the Minnesota and/or Des Moines rivers. *Notropis stramineus* is a creek and lake species that usually lives over a sand bottom. Only stragglers are found in large streams such as the Missouri, and it appears that this river presently constitutes a sufficient barrier to permit the two subspecies to maintain their identity despite close geographic proximity. Elsewhere, for example in northern Missouri from the Chariton River west and in southeastern Nebraska, there exist intergrading populations.

Recent impoundment of the Missouri River may produce habitat conditions suitable to *Notropis stramineus,* and thereby remove the barrier to mixing of the subspecies.

Vertebral counts (including 4 elements in the Weberian complex) were taken for one sample of each subspecies: *Notropis stramineus stramineus* from Whetstone Creek (station 8): 34 vertebrae (1 specimen), 35 (13),

36 (7), and 37 (1), N = 22, mean 35.36; *N. stramineus missuriensis,* from the Cheyenne River at Oral (station 104), 34 (2), 35 (4), 36 (8), 37 (3), N = 17, mean 35.71. The difference is small and probably of little or no taxonomic significance.

41a. *Notropis stramineus stramineus* (Cope)

Notropis deliciosus (misidentifications). Meek, 1892: 246 (Big Sioux R. [Sioux Falls, S. D., or Sioux City, Iowa]). Woolman, 1896: 351, 356 (Little Minnesota R., Browns Valley, Minn. [and S. D.]; Big Stone L., Ortonville, Minn. [and S. D.]; Wheatstone [Whetstone] Cr., Milbank). Cleary, 1956: 300 (Big Sioux R.; Lyon [Lincoln] and Sioux [Lincoln and Union] cos., Iowa [and S. D.]). Underhill, 1959: 98, 101 (Vermillion R.).

Notropis blennius (misidentification). Evermann and Cox, 1896: 402–03 (description, variation; James R., Mitchell).

Hybopsis blennius (misidentifications) .Churchill and Over, 1933: 50, fig. 37 (description, ecology, importance; streams and lakes east of Missouri R.).

It has been contended that the sand shiner comprises two subspecies in addition to *missuriensis,* both of which are said to occur in Indiana. Gerking (1945: 63) presented quantitative data for that state that fail to support the subspecific partition of eastern populations, and Bailey (1956: 332–33) observed that the separation is not adequately demonstrated. Nevertheless, the recognition of two eastern subspecies has been continued (Hubbs and Lagler, 1958: 77, 84); the recommended distinguishing characters are the same as those analyzed by Gerking and largely restate those proposed by Hubbs (1926: 37, 43) at a time when *Notropis volucellus* was regarded as a synonym of *Notropis deliciosus* auct. (= *stramineus*).

Notropis s. stramineus lives in creeks and lakes from the James River eastward; it has not been taken in the Missouri River or in the Red River drainage of South Dakota.

Station records: 3, 7, 8, 15, 16, 24, 26A, 26B, 26C, 27, 29, 35, 38, 39, 41, 43, 44, 45, 46, 49, 51.

41b. *Notropis stramineus missuriensis* (Cope)

Notropis deliciosus (misidentifications). Evermann, 1893b: 78 (Middle Cr. and Belle Fourche R. [Belle Fourche]; Rapid Cr. [Rapid City]; Cottonwood Cr. [Edgemont]; Hat Cr. [Ardmore]).

Notropis blennius (misidentifications). Evermann and Cox, 1896: 402–03 (description, variation; Belle Fourche R. and Middle Cr., Belle Fourche; Cheyenne R., Hot Springs; Cottonwood Cr., Edgemont; Hat Cr., Ardmore; Rapid Cr., Rapid City; Beaver Cr., Buffalo Gap; Crow Cr., [near] Chamberlain; Redwater Cr., Spearfish).

Hybopsis missuriensis Cope (1871: 437) was described from "near St. Joseph, Missouri," a locality that is interposed between the Missouri and

TABLE 5

Frequency Distribution of Body-Circumference Scale Counts in *Notropis stramineus* from South Dakota and Some Adjacent Areas

Drainage and Locality	S. D. Stations	Body-Circumference Scales																		No.	Mean	Percentage	
		20	21	22	23	24	25	26	27	28	29	30	31	32	33	34	35	36	37			20-25	26-37
Red River:																							
Sheyenne R., N.D.				1	7	17														25	23.6	100	0
Souris R., N.D.						15	13	4	2											34	24.8	82	18
Minnesota River:																							
Whetstone Cr., S.D.	7, 8		1	4	10	31	2	1												49	23.6	98	2
Little Sioux River:																							
Silver L., Ia.				5	16	18	3													42	23.5	100	0
Spirit L., Ia.				1		1	1													3	23.4	100	0
Meadow Cr., Ia.				1	6	7	2													16	23.6	100	0
Mill Cr., Ia.				2	5	25	5	3	2											42	24.2	88	12
Big Sioux River:																							
Brookings Co., S.D.	24, 25, 27, 29	1	2	14	38	113	31	15	4	1										219	24.0	91	9
Vermillion River:																							
Vermillion, S.D.	35			2	9	1	1													13	24.1	100	0
James River:																							
Jamestown, N.D.					1	19	17	11	2											50	24.4	74	26
Sanborn Co., S.D.	43				3	10	3	3	4	2										25	25.0	64	36
Yankton Co., S.D.	51				2	9														11	23.8	100	0
Missouri River:																							
Emanuel Cr., S.D.	53							4	3	6	9	1	1	1						25	28.3	0	100
Choteau Cr., S.D.	55, 56							3	7	20	11	3	4							48	28.3	0	100
Medicine Cr., S.D.	61									1	2		1							4	29.5	0	100
Bow R., Nebr.						1	2	1	2	9	6	7								28	28.2	11	89
Bazile Cr., Nebr.								1	4	6	6	4	1	2	1					26	29.2	0	100
Niobrara River:																							
Sheridan Co., Nebr.											3	5	6	3	3	6	3	2	1	32	32.5	0	100
Cherry Co., Nebr.									1		1	3	3	5			1			14	31.0	0	100
Snake R., Nebr.										1	2	3	1		1	1				9	30.4	0	100
Minnechaduza Cr., Nebr.							1		2	6	5	5	3	1	2					25	28.8	4	96
Antelope Cr., S.D.	84							1	1	9	2	5	1		1					20	28.8	0	100

TABLE 5 (continued)

FREQUENCY DISTRIBUTION OF BODY-CIRCUMFERENCE SCALE COUNTS IN *Notropis stramineus* FROM SOUTH DAKOTA AND SOME ADJACENT AREAS

Drainage and Locality	S. D. Stations	Body-Circumference Scales																		No.	Mean	Percentage	
		20	21	22	23	24	25	26	27	28	29	30	31	32	33	34	35	36	37			20-25	26-37
Missouri River:																							
Keyapaha R., Nebr.							1		1	9	4	2								17	28.2	6	94
White Clay Cr., S.D.	86									1	1	7	9	3	4					25	31.0	0	100
White R., S.D.	87									1	1	..	1							3	29.3	0	100
Little White R., S.D.	94					1	..	2	1	5	4	..	2							15	28.1	7	93
Bad R., S.D.	96									2	5	3	2	1						13	29.6	0	100
Cheyenne River:																							
Oral, S.D.	104									2	8	4	2	1	1					18	29.7	0	100
Rapid Cr., S.D.	116, 118							2	4	11	10	6	3	1						37	28.7	0	100
Wasta, S.D.	120							3	6	9	6	3	3	2						32	28.5	0	100
Missouri River:																							
Moreau R., S.D.	129, 130					1	3	2	4	10	4	1								25	27.4	16	84
S Fk. Grand R., S.D.	131							1	3	5	9	5	2							25	28.8	0	100
S Fk. Grand R., S.D.	132						1	1	6	7	4	5	1							25	28.2	4	96
Cedar R., N.D.								9	10	15	16	5	2	1						58	28.1	0	100
Cannonball R., N.D.										1	..	1								2	29.0	0	100
Heart R., N.D.									2	4	4	4	1	..	..	1				16	29.2	0	100
Little Missouri River:																							
Camp Crook, S.D.	136							2	4	4	6	6	2	1						25	28.8	0	100
Box Elder Cr., S.D.	137						1	6	3	2	2	1								15	27.1	7	93
Yellowstone River:																							
Pennel Cr., Mont.										3	6	9	2							20	29.5	0	100
Missouri River:																							
Nodaway R. dr., Mo.						17	15	16	6	7	1	1								63	25.6	51	49
Platte R. dr., Mo.				1	..	35	23	22	15	5	1	1								103	25.2	57	43
Totals:																							
N. s. stramineus		1	3	30	97	266	78	37	14	3										529	24.1	90	10
N. s. missuriensis						3	9	38	64	149	137	98	53	22	13	8	4	3	1	602	29.0	2	98
Intergrades (Mo.)				1	..	52	38	38	21	12	2	2								166	25.5	55	45

Platte rivers, Missouri, midway between the mouths of the Nodaway and Platte rivers. Since the species inhabits creeks it is unlikely that the specimen came from the Missouri River. Samples from the Platte and Nodaway drainages that presumably represent Cope's *missuriensis* prove to be intermediate between the subspecies (Table 5). In order to preserve conventional application of the name we restrict *Hybopsis missuriensis* Cope to the Plains subspecies, that is, the form with a high body-circumference scale count. *Hybopsis scylla* Cope, *Cliola chlora* Jordan, and *Hybopsis montana* Meek are synonyms of *Notropis stramineus missuriensis* (Cope).

Notropis stramineus missuriensis lives in all major drainages of western South Dakota. It has been taken once in Hipple Lake, a backwater of the Missouri River (sta. 66), and a single specimen (UMMZ 178861; body-circumference scales 30) has been received from Oahe Reservoir, 2 mi. above the dam. These are the only South Dakota records from the Missouri River. East of the Missouri it has been taken in Missouri River tributaries from Emanuel Creek (sta. 53) to Medicine Creek (sta. 61).

Station records: 53, 55, 56, 60, 61, 66, 84, 86, 87, 88, 94, 96, 97, 99, 104, 116, 117, 118, 120, 121, 127, 128, 129, 130, 131, 132, 136, 137.

42. *Notropis topeka* Gilbert—Topeka shiner
(Fig. 6)

Notropis topeka. Meek, 1892: 246 (Big Sioux R., Sioux City, Iowa [and S. D.]). Evermann and Cox, 1896: 403–04 (description, habitat; Firesteel*, Enemy, and Rock crs., Mitchell; Prairie Cr., Scotland). Underhill, 1959: 100 (upper Vermillion R.).

Codoma topeka. Churchill and Over, 1933: 53 (description, ecology, importance; small creeks of eastern and southern part of state).

The general distribution of *Notropis topeka* embraces parts of the Mississippi, Missouri, and Arkansas river drainages from southern Minnesota and southeastern South Dakota to central Missouri and Kansas (Fig. 6). It was formerly common in the Big Sioux, Vermillion, and James river drainages of South Dakota, but is now rare.

The accompanying map includes the following:

Iowa. Records mapped by Cleary (1956: 301) and: Mud Cr., Larchwood Twp., Lyon Co. (UMMZ 135501). Cleary's record from Van Buren County is not plotted since it apparently is erroneous.

Kansas. Records mapped by Minckley and Cross (1959: 211), also: Rock Cr., 9 mi. SW Lawrence (UMMZ 63162). Rock Cr., 12.5 mi S, 8.5 mi. E Topeka, Douglas Co. (UMMZ 156741). Topeka (UMMZ 86880). Near Saline [Salina] (UMMZ 86976). Fort Hays [Big Cr., Hays, Ellis Co.] (UMMZ 87193). Stevens Cr., trib. to Jacobs Cr., Cottonwood system [Neosho Drainage], 16 mi. SW Emporia, Oct. 11, 1936, and Mar. 1, 1937, Breukelman (UMMZ 120405, 120826). Upper Mill Cr. or Hendricks Cr., 0.5 mi. N Alma, Wabaunsee Co. (UMMZ 120835). 1 mi. N Alma (UMMZ 121915). Kansas R., Doc Wagner's farm, Riley

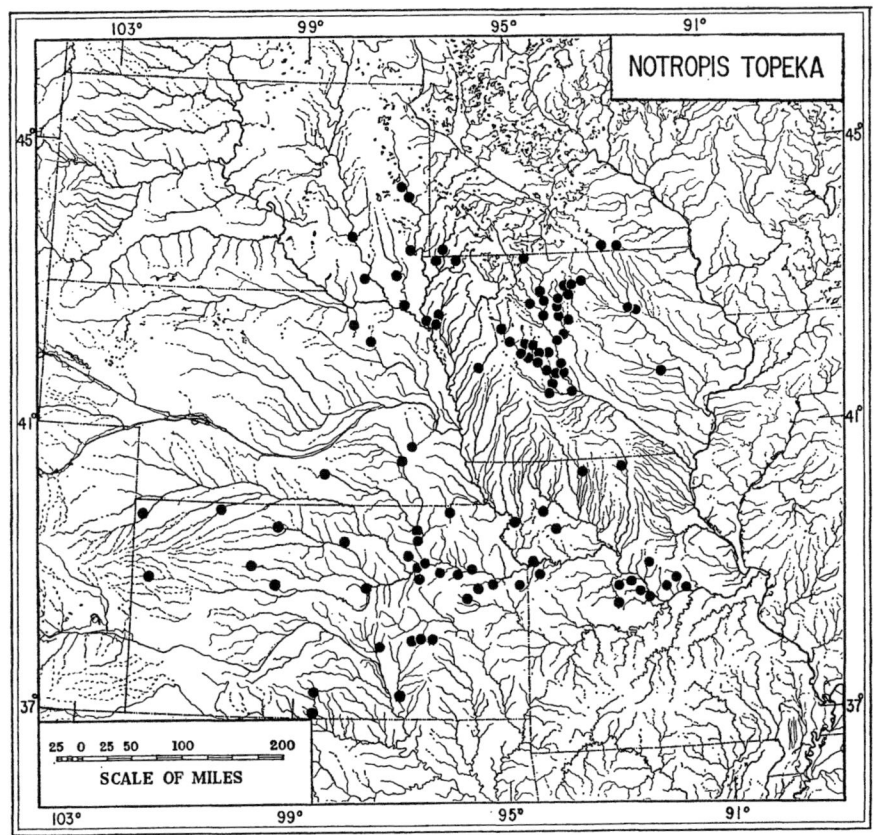

Fig. 6. Distribution by record stations of *Notropis topeka*.

Co. (UMMZ 121916). Gamefork Cr., 1.5 mi. W Irving, Marshall Co. (UMMZ 121917). Gamefork Cr., Marshall Co. (UMMZ 122068). Wildcat Cr., 3.5 mi. from Riley, Riley Co. (UMMZ 121918). Deep Cr., Pillsbury Crossing near Manhattan, Riley Co. (UMMZ 121998; 128811; 144913). Creek S Osburne, Riley Co. (UMMZ 122066). Nemaha Cr., below Nemaha State Lake, Nemaha Co. (UMMZ 144866). Old bed Wildcat Cr., 2 mi. SW Manhattan, Riley Co. (UMMZ 144883). Trib. to Clarks Cr., 2.5 mi. W, 2 mi. N Dwight, Geary Co. (UMMZ 160488). Tomahawk Cr., 2 mi. N, 0.5 mi. W Stanley, Johnson Co. (UMMZ 160553).

Minnesota. Records cited by Underhill (1957: 19), also: Creek, 16 mi. E Austin, Mower Co. (UMMZ 127672).

Missouri. Branch of Big Tavern Cr., 3 mi. NE Portland, Galloway Co. (UMMZ 148000). Silver Fork, Roche Perch Cr., 11 mi. N Columbia, Boone Co. (UMMZ 148029). Lost Cr., 2 mi. SW Weatherby, Dekalb Co. (UMMZ 148163). Shoal Cr., 3 mi. W Livonia, Putnam Co. (UMMZ 148753). East Muddy Cr., 2 mi. NE Princeton, Mercer Co. (UMMZ 148797). Big Shoal Cr., 0.5 mi. SE Ravena Gardens on U. S. 69, Clay Co. (UMMZ 148845). Shoal Cr., 0.5 mi. N Kingston, Caldwell Co. (UMMZ 148905). Crows Cr., 3 mi. NE Fulton,

Callaway Co. (UMMZ 149312). Middle R., 4 mi. SW Fulton, Callaway Co. (UMMZ 149300). Brush Cr., 8 mi NE California, Moniteau Co. (UMMZ 150556). Clarks Fork Cr., 5 mi. SE Boonville, Cooper Co. (UMMZ 150595). Petite Saline Cr., 4¼ mi. W Bunceton, Cooper Co. (UMMZ 150628). Little Blue R., 5 mi. NW Lees Summit, Jackson Co. (UMMZ 152565). Rock Cr., 5 mi. NE Centertown, Cole Co. (UMMZ 152656). Richland Cr., 1 mi. E Florence, Morgan Co. (U. Mo.).

Nebraska. Records by Meek (1894: 136) from Salt Cr., Lincoln, and Blue R., Crete. Pond at Creighton (UMMZ 86403). North Branch, 3 mi. N Pierce, Pierce Co. (UMMZ 134035). Little Blue R., Ayr Reservation Grounds, 1 mi. N, 0.5 mi. W Ayr, Adams Co. (UMMZ 134266; 134988).

South Dakota. Station records: 23, 26D, 32, 34, 35.

43. *Notropis heterolepis* Eigenmann and Eigenmann—Blacknose shiner

Notropis cayuga. Meek, 1892: 246 (Big Sioux R., Sioux City, Iowa [and S. D.]). Evermann and Cox, 1896: 401–02 (description, habitat; James R., Enemy, Firesteel, and Rock crs., Mitchell; Prairie Cr.*, Scotland; Choteau and Emanuel crs., Springfield). Woolman, 1896: 351 (description; Little Minnesota R., Browns Valley, Minn. [and S. D.]).

Hybopsis cayuga. Churchill and Over, 1933: 48, fig. 34 (description, ecology, importance; creeks and small rivers east of Missouri R.).

Notropis heterolepis. Moyle and Clothier, 1959: 179 (Lake Traverse).

The blacknose shiner was once one of the common cyprinids of eastern South Dakota, but it is now rare. We have not collected it, and have examined only the Prairie Creek specimens reported by Evermann and Cox. *Notropis heterolepis* lives in clear, vegetated waters. As a result of drought and intensive land use this habitat has been greatly restricted during recent years throughout the Plains and Prairie regions from South Dakota to Ohio (Cleary, 1956: 272, 301; Trautman, 1957: 388–390). In contrast, the species is widespread and abundant in areas with clear, sluggish or standing water, such as Minnesota (Underhill, 1957: map 9), Wisconsin (Greene, 1935: 96), and Michigan.

KEY TO SPECIES OF *Hybognathus*

1a.—Body yellowish in life. Fins more rounded. Scales with the radii numerous (usually nearly 20 in adult) and weak; circuli smoothly curved at basal corners of scale. Head blunter. Size smaller, length to about 4 inches............................
.... Brassy minnow, *Hybognathus hankinsoni*
1b.—Body silvery in life. Fins higher. Scales with the radii few (about 10) and strong; circuli sharply angulate (more or less squared) at basal corners of scale. Head more elongate. Size larger, length to about 6 inches. 2
2a.—Scale rows across belly 15 to 22, usually 16 to 18 (counted just in advance of pelvic fins, not including lateral-line rows). Inferior retractor muscles of pharyngeal arches closely approximated posteriorly (Pl. I); their origins on the posterior end of the depressed, narrow basioccipital process, which is not expanded posteriorly. Width of process 1.7 to 5.0 in its length. Eye smaller (see Fig. 7)........................
.... Plains minnow, *Hybognathus placitus*

2b.—Scale rows across belly 11 to 17, usually 12 to 15. Inferior retractor muscles of pharyngeal arches well separated posteriorly (Pl. I); their origins on the ends of lateral posterior expansions of the much depressed basioccipital process. Width of process 0.9 to 1.5 in its length. Eye larger (see Fig. 7).............................
............................Silvery minnow, *Hybognathus nuchalis nuchalis*

44. *Hybognathus hankinsoni* Hubbs—Brassy minnow

Hybognathus nubilum (misidentifications). Evermann and Cox, 1896: 397 (comparison; Emanuel and Choteau crs., Springfield; Crow Cr.* and White R., [near] Chamberlain).

Dionda nubila (misidentification). Churchill and Over, 1933: 41 (description; streams east of Missouri R.).

Hybognathus hankinsoni. Bailey, 1954: 290–91, fig. 1 (description; distribution, map). Underhill, 1957: map 21 (Lake Traverse and Little Minnesota R. [Roberts Co.]). Hugghins, 1959: 22 (Medary Cr.). Underhill, 1959: 101 (Vermillion R.).

The brassy minnow is a common creek and small-river fish that is widespread in South Dakota; collections have been taken in all major drainages except the Missouri River itself and those western streams from the Cheyenne to North Dakota.

Station records: 1, 3, 5, 8, 9, 25B, 26C, 27, 28, 29, 34, 39, 41, 42, 43, 44, 48, 49, 53, 55, 56, 59, 84, 136.

45. *Hybognathus placitus* Girard—Plains minnow
(Pl. I; A, B)

Hybognathus evansi. Girard, 1856: 18 (in part?; original description; [Missouri R.] Ft. Pierre, "Nebraska"). Girard, 1858: 236 (in part?; description; [Missouri R.] Ft. Pierre, "Nebraska"). Cope, 1879: 440 (in part?; characters; Battle [? = Blue Blanket] Cr.).

Hybognathus nuchale evansi. Evermann and Cox, 1896: 396–97 (in part; description; Cheyenne R.*, Hot Springs and Edgemont; Cottonwood Cr.*, Edgemont; Hat Cr., Ardmore; Belle Fourche R. and Middle Cr., Belle Fourche).

Hybognathus nuchalis placita. Evermann, 1893b: 78 ("S. Fork of" Cheyenne R.* [Hot Springs and Edgemont]; Cottonwood Cr.* [Edgemont]; Middle Cr. and Belle Fourche R. [Belle Fourche]; Hat Cr. [Ardmore]).

Hybognathus churchilli. Hildebrand, 1932; 257–60, fig. 1 (original description, variation, comparison; Cheyenne R. near mouth of Cherry Cr., holotype, USNM 92248; White River, near White River; Bad R., near Midland). Churchill and Over, 1933: 41, fig. 25a (description, habitat; Cheyenne, White, and Bad rivers).

Hybognathus nuchalis (misidentifications). Churchill and Over, 1933: 40–41, fig. 25 (in part; description, ecology, importance; rivers west of Missouri R.).

Hybognathus nuchalis nuchalis (misidentification). Bailey, 1956: 333 (in part; synonymy; Little Missouri R., Camp Crook [Camp Creek an error], Harding Co.).

Based chiefly on study of a large collection of *Hybognathus* taken in the Little Missouri River at Camp Crook, South Dakota (station 136), Bailey (1956: 333) synonymized *H. placitus* with *H. n. nuchalis*. He noted the differences employed to separate these forms, but because in this collection they characterized fish from separate age classes he attributed them to environmental modification. Since then Dr. George A. Moore and Mr. A. D. Niazi, Oklahoma State University, in a study of the Weberian apparatus of these forms (MS), and Dr. Teruya Uyeno, University of Michigan, while studying North American cyprinids, have independently discovered a remarkable difference between these species in the configuration of the posterior process of the basioccipital (Pl. I). In *Hybognathus* the process is typically depressed and flat, with or without a perceptible median keel. In *H. placitus* the process is relatively slender and bladelike, without developed lateral expansions posteriorly; the ratio of greatest width in length behind the pharyngeal pad varies from 1.7 to 5.0. The process functions mechanically as attachment for the retractor arcus branchialis dorsalis inferior muscle (Takahasi, 1925), which originates from the ventrolateral surface near the posterior end. The muscles of the two sides are closely approximated at their origins so that in ventral aspect they are V-shaped. In *H. nuchalis* the process is expanded posteriorly and the posterior margin is either truncate (shovel-shaped) or emarginate. The lateral angles either are flat or are deflected ventrally and to them are attached the retractor muscles, which are therefore well separated from one another. The basioccipital process has a width-in-length ratio of 0.9 to 1.5. The observed structural differences are not sex correlated and apparently do not vary significantly with growth. With rare exceptions, determinations can be made quickly and with confidence. It is apparent that *H. placitus* and *H. nuchalis* are distinct species which are in part sympatric; whether or not hybridization occurs awaits further study.

Re-examination of the large collection from station 136 with the aid of the newly discovered difference confirms identification of large fish (1950 year class) as *H. nuchalis* and of most smaller fish (1951 and 1952 year classes) as *H. placitus*. Three young fish, however, prove to be *nuchalis*. Presumably some unknown behavioral or distributional difference between the species is responsible for the aberrant population structure here.

External differences that have been employed previously to distinguish *Hybognathus placitus* from *H. nuchalis* are not sufficiently marked to permit consistently accurate determinations, especially for juveniles. Eye size is commonly used, and rather high accuracy in separation of the forms can be achieved by a skilled taxonomist thoroughly familiar with the species. Nevertheless, the character overlaps, is subject to allometric change (Fig. 7), and probably is influenced by environmental factors. One of the most useful characters involves the scales on the belly; in *placitus* these average smaller and are more irregular. Many individuals can be determined

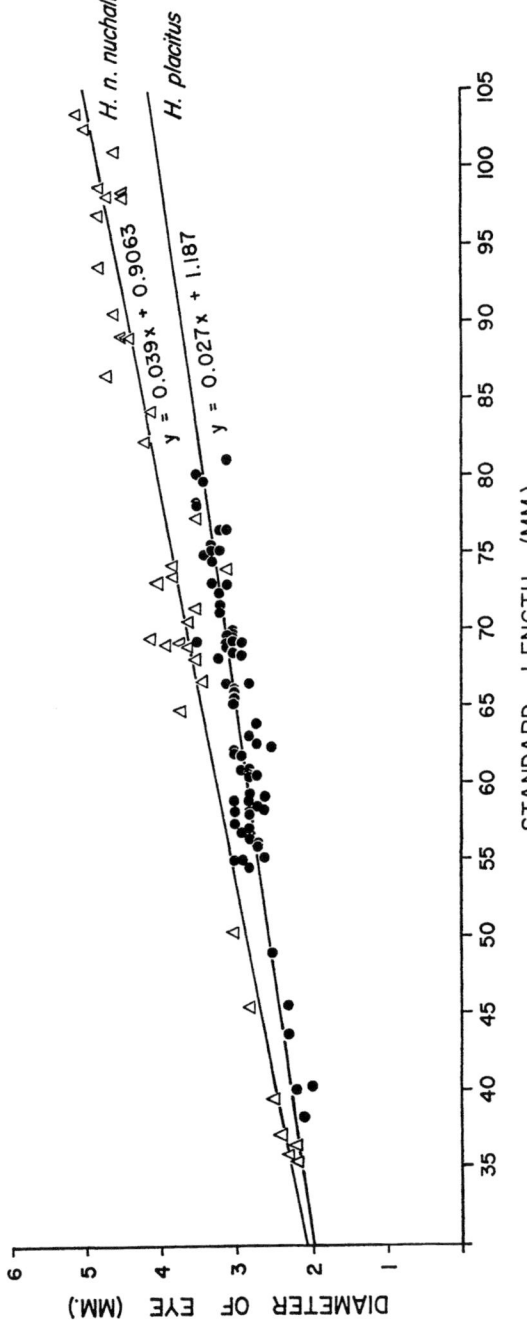

Fig. 7. Regression of eye size on standard length in *Hybognathus placitus* and *H. n. nuchalis*. Data from two collections in the Little Missouri River, North Dakota, where the species were collected sympatrically: *H. placitus* UMMZ 94799, at Marmarth, and No. 94804, below Marmarth; *H. n. nuchalis* No. 94801, Marmarth, and No. 94800, below Marmarth.

accurately at a glance or by a count of transverse scales across the belly between the lateral lines (Table 6). But the species overlap in this character, as determined by repeated dissections to reveal the pharyngeal process. Where the species are sympatric, *placitus* always averages more scales. Populations vary considerably in this count, and further extension of data as well as continued search for additional characters is needed to facilitate identification.

Vertebral counts of *H. placitus* from the Little Missouri River, station 136, are 37 (4), 38 (30), 39 (31), 40 (3); mean for 68 specimens, 38.49; S.D., 0.68; S.E., 0.08.

In South Dakota *Hybognathus placitus* reaches its maximum abundance in the smaller Plains streams of the western part of the state. At such locations it commonly occurs to the exclusion of *H. nuchalis*. In the middle and lower courses of western streams the species is associated with or replaced by *nuchalis*. In the Missouri River the latter is much the commoner species. We have no verified record of *H. placitus* from eastern or northern tributaries to the Missouri.

Station records: 65, 66, 72, 76, 79, 83, 87, 88, 91, 96, 97, 99, 100, 102, 104, 120, 127, 128, 129, 130, 131, 132, 133, 136, 137.

46. *Hybognathus nuchalis nuchalis* Agassiz—Silvery minnow
(Pl. I; C, D)

Hybognathus evansi. Girard, 1856: 18 (in part?; original description; [Missouri R.] Ft. Pierre, "Nebraska"). Girard, 1858: 236 (in part?; description; [Missouri R.] Ft. Pierre, "Nebraska"). Cope, 1879: 440 (in part?; characters; Battle [? = Blue Blanket] Cr.).

Hybognathus nuchale evansi. Evermann and Cox, 1896: 396–97 (in part; description; Choteau and Emanuel* crs., Springfield; Crow Cr. and White R., [near] Chamberlain).

Hybognathus nuchalis. Meek, 1892: 245–46 (Big Sioux and Missouri rivers, Sioux City, Iowa [and S. D.]). Hildebrand, 1932: 260 (characters, comparison; Cheyenne R., near mouth of Cherry Cr.; White R., near White R.; Bad R., Midland). Churchill and Over, 1933: 40–41, fig. 25 (in part; description, ecology, importance; rivers both east and west of Missouri R.).

Hybognathus nuchalis nuchalis. Cleary, 1956: 303 (Big Sioux R., Lyon [Lincoln], Sioux [Lincoln and Union], and Plymouth [Union] cos., Iowa [and S. D.]). Bailey, 1956: 333 (in part; synonymy; Little Missouri R.*, Camp Crook [Camp Creek an error], Harding Co.). Underhill, 1959: 101 (lower Vermillion R.).

Because of the close superficial similarity of this species and *H. placitus* (see p. 72 and Table 6), and their past confusion, much of the synonymy is unclear. We have not re-examined the type materials of *H. evansi* which may consist of either or both species. Vertebral counts of *H. n. nuchalis* from the Little Missouri River, station 136, are 38 (7), 39 (19), 40 (3), 41 (3); mean for 32 specimens, 39.06; S.D., 0.84; S.E., 0.15.

TABLE 6
Frequency Distribution of Scale Rows Across Belly in *Hybognathus placitus* and *H. n. nuchalis* from South Dakota

Species and Locality (Station or UMMZ No.)	Scale Rows Across Belly												N	M
	11	12	13	14	15	16	17	18	19	20	21	22		
H. placitus														
Missouri R. (76)					1								1	
" " (83)							2	8	7				17	18.3
White R. (87)						2	..	..	1				3	17.0
" " (88)			2	1	2	5	7	7	3				27	16.7
" " (91)										1			1	
Bad R. (96)							1						1	
Cheyenne R. (104)							1	1	3	1			6	18.7
" " (120)									2	1	1		4	19.8
" " (86767)					1	1	5	1	1				9	17.0
Belle Fourche Res. (128)							1						1	
Moreau R. (129)									1	..	1		2	20.0
" " (130)								1	1	..	2		4	19.8
S Fk. Grand R. (131)					1	4	5	8	..	1	1		20	18.4
" " " " (132)					3	3	8	7	2				23	17.1
Little Missouri R. (136)					4	10	26	21	15	6			82	17.6
Box Elder Cr. (137)							2	2	..	2	1		7	18.7
H. n. nuchalis														
Emanuel Cr. (86769)		1											1	
Medicine Cr. (61)					1	2	2						5	16.2
Missouri R. (64)					1	2	2						5	16.2
" " (67)						1							1	
" " (76)			3	12	4	1							20	14.2
" " (83)			10	10	23	7							50	14.5
White R. (88)	1	1	..	1	3	1							7	14.0
Moreau R. (129)				1	5								6	14.8
" " (130)	1	..	1	1	3	2							8	14.4
Grand R. (135)			3	11	1								15	13.9
Little Missouri R. (136)		1	1	11	18	2							33	14.6
Totals														
H. placitus			2	1	11	22	57	53	44	11	6	1	208	17.7
H. n. nuchalis	2	2	19	47	59	17	5						151	14.9

On the basis of South Dakota material examined, *H. nuchalis* appears to live to the exclusion of *placitus* in the lower reaches of some eastern and northern tributaries to the Missouri, to be sympatric with but to outnumber *placitus* in the Missouri River and in the lower courses of larger western tributaries, but to be largely or entirely replaced by *placitus* in the smaller western Plains streams.

Station records: 61, 64, 65, 67, 72, 73, 76, 83, 88, 130, 135, 136.

KEY TO SPECIES OF *Pimephales*

1a.—Lateral line complete. Mouth almost horizontal, subterminal. Body slender and terete. Caudal spot prominent. No nuptial tubercles on mandible in breeding males, which have a barbellike flap at angle of mouth. ....Bluntnose minnow, *Pimephales notatus*
1b.—Lateral line incomplete. Mouth strongly oblique, terminal. Body compressed and deeper. Caudal spot faint. Nuptial tubercles present on mandible and snout in breeding males, which have no barbellike flap of skin. ....
.... Fathead minnow, *Pimephales promelas*

47. *Pimephales notatus* (Rafinesque) —Bluntnose minnow

Pimephales notatus. Meek, 1892: 246 (Big Sioux R., Sioux Falls, S. D., and Sioux City, Iowa [and S. D.]). Woolman, 1896: 348, 351 (characters; Daugherty Cr., Browns Valley, Minn. [and S. D.]; Big Stone L., Creagers farm and Ortonville, Minn. [and S. D.]). Cleary, 1956: 304 (Big Sioux R., Lyon [Lincoln] and Sioux [Lincoln and Union] cos., Iowa [and S. D.]).
Ceratichthys vigilax (misidentification). Churchill and Over, 1933: 42–43, fig. 27 (description, ecology, importance; smaller rivers).

In South Dakota the bluntnose minnow is known only in the Red, Minnesota, and Big Sioux river drainages, and, presumably as an introduction, in Flat Creek Lake near Shadehill Reservoir on the Grand River, Perkins County (Sta. 134). In North Dakota it occurs also in the James River (Woolman, 1896: 359). In Nebraska Johnson (MS) found the bluntnose minnow only in the Elkhorn River system, where it is native (Meek, 1894: 136). In Iowa the species is common in the Big Sioux, Floyd, and Little Sioux drainages (Cleary, 1956: 304); elsewhere in the Missouri drainage it is rare and perhaps occurs only as a result of introductions from the Des Moines basin, where it abounds. It seems evident that the species entered the middle Missouri drainage across the divide from the upper Mississippi (pp. 122–25). It also remained in the lower part of the Missouri system during the last glaciation or entered it in postglacial time. The Elkhorn stock could conceivably have come from either source or may be a glacial relict.

Station records: 3, 4, 5, 7, 8, 9, 24, 26A, 26C, 29, 134.

48. *Pimephales promelas* Rafinesque—Fathead minnow

Hyborhynchus nigellus. Cope, 1879: 440–41 (Battle Cr. [? = Blue Blanket Cr.]).
Pimephales promelas. Evermann, 1893b: 78 (Middle Cr. [Belle Fourche], Rapid Cr. [Rapid City], Cottonwood Cr. [Edgemont], and Hat Cr. [Ardmore]). Evermann and Cox, 1896: 397–99 (description, habitat; James R., Enemy and Rock crs., Mitchell; Choteau and Emanuel* crs., Springfield; Crow Cr., [near] Chamberlain; Rapid Cr., Rapid City; Hat Cr., Ardmore; Cottonwood Cr., Edgemont; Middle Cr., Belle Fourche; French Cr., Custer). Churchill and Over, 1933: 42, fig. 26 (description, ecology, importance; streams east of Missouri R.). Cleary, 1956: 304 (Big Sioux R., Lyon [Lincoln] and Sioux [Lincoln and Union] cos., Iowa [and S. D.]). Underhill, 1957: map 22 (Lake Traverse and Little Minnesota R. [Roberts Co.]). Hugghins, 1959: 22 (pond near Waubay, Hidewood Cr., Big Sioux R.). Moyle and Clothier, 1959: 179 (Lake Traverse). Underhill, 1959: 101 (Vermillion R. drainage).
Cliola smithii. Evermann and Cox, 1896: 400–01 (original description; Prairie Cr., Scotland).
Ceratichthys smithii. Churchill and Over, 1933: 43 (compiled, description; Prairie Cr., Scotland).

The fathead minnow is probably the most ubiquitous fish in South Dakota, and is one of the commonest bait minnows. It occurs in all major drainages, is ecologically adaptable, and is usually common where found. It is uncommon in the Missouri River proper and is absent from the higher streams of the Black Hills.

Station records: 1, 2, 3, 4, 5, 8, 9, 10, 11, 12, 13, 14, 15, 16, 17, 18, 19, 20, 21, 22, 23, 24, 25B, 26A, 26B, 26C, 27, 28, 29, 30, 31, 33, 34, 35, 36, 38, 39, 41, 43, 44, 46, 48, 49, 53, 55, 56, 57, 58, 59, 60, 61, 76, 84, 85, 86, 87, 88, 89, 90, 92, 93, 95, 96, 97, 98, 100, 104, 111, 118, 121, 127, 129, 130, 131, 132, 133, 135, 136, 137.

49. *Campostoma anomalum* (Rafinesque)—Stoneroller

Campostoma anomalum. Evermann and Cox, 1896: 395 (characters; Emanuel Cr., Springfield; Enemy* and Firesteel* crs., Mitchell; Prairie Cr., Scotland; Crow Cr., [near] Chamberlain). Woolman, 1896: 356 (habitat; Wheatstone [Whetstone] Cr., Milbank). Churchill and Over, 1933: 39–40, fig. 24 (description, ecology, importance; small rivers and creeks). Cleary, 1956: 305 (Big Sioux R., Lyon [Lincoln] and Sioux [Lincoln and Union] cos., Iowa [and S. D.]). Underhill, 1957: map 24 (Little Minnesota R. [Roberts Co.] and S Fork of Yellowbank R. [Grant Co.]). Hugghins, 1959: 22 (Medary Cr.). Underhill, 1959: 101 (Vermillion R.).

Except for the Red River drainage, where it is absent, the stoneroller is common to all hydrographic basins east of the Missouri River. Elsewhere in South Dakota it is known to occur only in the Niobrara and White river drainages.

Station records: 3, 7, 8, 9, 11, 16, 19, 25B, 26B, 27, 28, 29, 41, 42, 43, 84, 93.

CATOSTOMIDAE—SUCKERS

KEY TO GENERA OF CATOSTOMIDAE

1a.—Dorsal fin longer, with more than 20 principal rays. Supraorbital bone present. ... 2
2a.—Lateral-line scales more than 50. Lips papillose. Head small, abruptly more slender than body. Eye closer to back of head than to tip of snout. .... *Cycleptus*
2b.—Lateral-line scales fewer than 50. Lips smooth or weakly plicate. Head larger and not abruptly more slender than body. Eye closer to tip of snout than to back of head. 3
3a.—Cheek shallow and shortened (distance from eye to lower posterior angle of preopercle about three-fourths that to upper corner of gill-cleft). Subopercle broadest at middle, subsemicircular. Mouth terminal to inferior. Anterior fontanelle much reduced or obliterated. .... *Ictiobus*
3b.—Cheek relatively deep and long (eye about equidistant from upper corner of gill-cleft and posteroventral angle of preopercle). Subopercle broadest below its middle, subtriangular. Mouth inferior. Anterior fontanelle well developed. .....
.... *Carpiodes*
1b.—Dorsal fin shorter, with 17 or fewer principal rays. No supraorbital bone. 4
4a.—Lateral line with 50 or fewer scales 5
5a.—Head depressed between eyes, the interorbital area concave. Lips heavily papillose. Swimbladder with 2 chambers. Caudal-peduncle scales usually 16. ..*Hypentelium*
5b.—Head not depressed between eyes, the interorbital area flat or convex. Lips plicate or weakly papillose (in *anisurum*). Swimbladder with 3 chambers. Caudal-peduncle scales usually 12. .... *Moxostoma*
4b.—Lateral line with more than 55 scales. (Lips heavily papillose. Swimbladder with 2 chambers.) 6
6a.—Posterolateral corners of lips smoothly rounded; halves of lower lip almost separated by a deep median incision. Mouth with the cartilaginous ridges scarcely to moderately developed, that of lower jaw semicircular. Gut with several folds, but not extensively coiled; its length about 1.5 to 4.0 times standard length. Head terete, its width about equal to or little greater than its depth. .... *Catostomus*
6b.—Posterolateral corners of lips each with a deep notch; halves of lower lip broadly conjoined anteriorly, the bridge bearing 2 or more rows of papillae. Cartilaginous cutting ridges of mouth well developed, that of lower jaw truncate. Gut extensively coiled, its length about 6 times standard length. Head depressed, its width much greater than its depth. .... *Pantosteus*

50. *Cycleptus elongatus* (LeSueur)—Blue sucker

Cycleptus elongatus. Harrison and Speaker, 1954: 515 (Big Sioux R., Iowa [and S. D.]). Cleary, 1956: 283 (Big Sioux R., Plymouth [Union] Co., Iowa [and S. D.]). Underhill, 1959: 100 (Missouri R., at mouth of Vermillion R.).

Except for the reported occurrence in the Big Sioux River, we know the blue sucker in South Dakota only from the Missouri River. It is taken regularly but in moderate numbers in Missouri River impoundments and their tailwaters. We took the species but once, over exposed shale bedrock, an unusual bottom type in the river. This is perhaps a preferred or limiting habitat.

Station records: 70, 72.

KEY TO SPECIES OF *Ictiobus*

1a.—Mouth large and oblique; upper lip about level with lower margin of orbit; upper jaw about as long as snout. Lips thin, only faintly striate. Lower pharyngeal arch thin, more than twice as high as wide. .... Bigmouth buffalo, *Ictiobus cyprinellus*
1b.—Mouth smaller, little oblique; upper lip far below lower margin of orbit; upper jaw distinctly shorter than snout. Lips fuller, more or less coarsely striate. Lower pharyngeal arch heavy, about as wide as high. ... Smallmouth buffalo, *Ictiobus bubalus*

51. *Ictiobus cyprinellus* (Valenciennes)—Bigmouth buffalo

Ictiobus cyprinella. Cox, 1896: 608 (presumptive report; Big Stone L. [Roberts Co.]).
Megastomatobus cyprinella. Churchill and Over, 1933: 31, fig. 16 (description, ecology, importance; streams and lakes east of Missouri R.). Shields, 1958a: 31–32; 1958b: 360 (Fort Randall Res.).
Ictiobus cyprinellus. Cleary, 1956: 284 (Big Sioux R., Lyon [Lincoln], Sioux [Lincoln], and Woodbury [Union] cos., Iowa [and S. D.]). Allum and Hugghins, 1959: 34 (Lake Poinsett, Brookings and Hamlin cos.). Hugghins, 1959: 21 (Big Stone L., L. Poinsett, Goldsmith L., L. Hendricks).

The bigmouth buffalo is moderately common in larger streams and lakes of eastern South Dakota. In the Missouri River the species was present in small numbers prior to impoundment, but an increase may be anticipated since the slower and clearer waters associated with reservoir construction should favor it.

Recent information from Mr. J. T. Shields indicates that nearly one-half million pounds of buffaloes, largely this species, were harvested commercially in Fort Randall Reservoir in the fiscal year 1960–1961.

Station records: 8, 24, 35, 38, 39, 51, 72, 83.

52. *Ictiobus bubalus* (Rafinesque)—Smallmouth buffalo

Ictiobus bubalus. Churchill and Over, 1933: 32, fig. 17 (description, ecology; [streams and lakes east of Missouri R.]). Shields, 1958a: 31–32; 1958b: 360 (Fort Randall Res.). Underhill, 1959: 100 (lower Vermillion R.).
Ictiobus niger (probable misidentifications). Cleary, 1956: 284 (Big Sioux R., Lyon [Lincoln], Sioux [Union], and Plymouth [Union] cos., Iowa [and S. D.]).

The distinction between *Ictiobus bubalus* and *Ictiobus niger* is fine and the relationship of the two is in need of more thorough study. In some waters they usually can be separated; in others they seem to grade into one another. Whatever the systematic relationship, we know of no evidence that the black buffalo occurs in South Dakota, except for Cleary's records, and he did not map the smallmouth buffalo from the Big Sioux or Missouri rivers.

The smallmouth buffalo is known to occur in South Dakota only in the Missouri River, where taken during recent years by fishery biologists in the Oahe (UMMZ 178862) and Fort Randall reservoirs, and in the lower courses of the James, Vermillion, and Big Sioux rivers.

Station record: 51.

KEY TO SPECIES OF *Carpiodes*

1a.—Scales smaller, in 37 to 40 rows along body. Lower lip without trace of a median, nipplelike projection. Opercular striations weak in adults, scarcely evident in young. Tip of lower lip clearly in advance of anterior nostril. Snout produced; distance from its tip to anterior nostril equal to length of eye (much greater than eye in adults).
.... Quillback, *Carpiodes cyprinus*

1b.—Scales larger, in 33 to 36 (occasionally 37) rows along body. Lower lip with an evident median, nipplelike projection. Opercle strongly striated in adults (weakly striate in young). Tip of lower lip scarcely or not at all in advance of anterior nostril. Snout blunter; distance from its tip to anterior nostril less than eye (equal in large adults).
.... River carpsucker, *Carpiodes carpio carpio*

53. *Carpiodes cyprinus* (LeSueur)—Quillback

Carpiodes forbesi. Bailey, 1951: 191 (Big Sioux R.*, Lyon [Lincoln] Co., Iowa [and S. D.]). Harrison and Speaker, 1954: 515 (Big Sioux R.*, Iowa [and S. D.]). Cleary, 1956: 74, 285 (description, life history; Big Sioux R., Lyon [Lincoln], Sioux [Lincoln and Union], and Woodbury [Union] cos., Iowa [and S. D.]).

Carpiodes cyprinus. Cleary, 1956: 286 (Big Sioux R., Lyon [Lincoln], Sioux [Union], and Woodbury [Union] cos., Iowa [and S. D.]).

Carpiodes cyprinus cyprinus. Trautman, 1956: 34, 39 (northeastern South Dakota).

During recent years most workers have recognized four species of *Carpiodes* from the United States (e.g., Forbes and Richardson, 1909; Hubbs, 1930; Bailey, 1956; Moore, 1957). These fall into two readily distinguishable groups, of which the *carpio-velifer* pair is discussed under *Carpiodes carpio*. Hubbs (1930) recognized that the names assigned by Forbes and Richardson (1909) to the members of the other pair were improper; he applied *Carpiodes cyprinus* to the quillback (their *C. velifer*) and proposed a new name, *Carpiodes forbesi*, for the other species: "I am not convinced that this species is valid, but I propose for it a name, for the sake of convenience and emphasis. That the name *thompsoni* is erroneously associated with it, is shown in the account of the following species [*C. cyprinus*] . . . ." The chief characters used to distinguish *forbesi* from *cyprinus* are the more slender form, the large head and mouth, and the lower and more posterior dorsal fin. *Carpiodes forbesi* has been reported, or specimens meeting its description have been collected, from southern Minnesota, Wyoming, Nebraska, northern Missouri, and Iowa through Illinois and Indiana to central Ohio. In discussing *Carpiodes cyprinus* in Ohio, Trautman (1957: 237) has noted that in large waters of low turbidity and with an abundance of food the fish are excessively fat, deep-bodied, and small-eyed, whereas "individuals from turbid waters containing little food, and others heavily parasitized, grow slowly, are terete and large-eyed and resemble *Carpiodes forbesi*, a form inhabiting turbid streams west of the Mississippi River." Trautman has reidentified Ohio specimens previously thought to be *forbesi* as *C. cyprinus*. *C. forbesi* is said to be chiefly a species of prairie and plains areas (Hubbs, 1945), a region where high turbidity and scanty food supplies in the rivers are characteristic. We submit that it is illogical to interpret the *"forbesi"* phenotype where it occurs east of the Mississippi as an emaciated variant of *cyprinus*, while assigning western specimens to a different species, *forbesi*, the type locality of which is in the Illinois River. Although one of us (Bailey) has previously recognized *forbesi* as a valid species, we now feel, as Trautman does for Ohio specimens, that the slender fish with a low dorsal are likely the product of their environment. It should be noted in analogy that in turbid plains and prairie streams *Cyprinus carpio* is commonly slender, though not emaciated, and of notably different form from fast-growing lacustrine fish. The problem deserves further study, but in 32 years since its qualified proposal insufficient evidence of the validity of *forbesi* has accumulated to justify its retention. We regard it as a modification of *cyprinus*.

The differences employed by Trautman (1956) to distinguish subspecies of *Carpiodes cyprinus* appear to result from similar but less drastic molding by extrinsic forces. The best evidence for this conclusion has been supplied by Trautman himself (1956: 38; 1957: 237). For *C. c. hinei*, the slender body, small eye, and large head contrast with *C. c. cyprinus*, but approach *forbesi*. These modifications correlate with the regionally intermediate environmental conditions that prevail in the range of *hinei* (Ohio and middle Mississippi basins) as contrasted with the clear-water, lacustrine northern subspecies, *C. c. cyprinus*, and with the turbid-water plains form, *C. forbesi*. It is probably not mere coincidence that these three nominal forms show strikingly close agreement in fin-ray and scale counts (Trautman, 1956: 37). Nor should it be surprising that in a lacustrine habitat, such as Buckeye Lake, Ohio, within the range of *hinei* the phenotype closely resembles *C. c. cyprinus*.

Carpiodes cyprinus is known in South Dakota from the Big Sioux River, where both the typical and the *"forbesi"* forms have been found (Cleary,

1956), and from the Minnesota River drainage, where only the typical form has been taken. This is the only species of *Carpiodes* known to have utilized the Warren Outlet into the Red River basin.

Station records: 7, 8.

54. *Carpiodes carpio carpio* (Rafinesque)—River carpsucker

Carpiodes damalis. Girard, 1858: 218–19, pl. 48, figs. 1–4 (description; [Missouri R.] Ft. Pierre, "Nebraska").

Carpiodes velifer (misidentifications). Meek, 1892: 245–46 (Missouri and Big Sioux* rivers, Sioux City, Iowa [and S. D.]). Cleary, 1956: 287 (Big Sioux R., Lyon Co., Iowa [and Lincoln Co., S. D.]).

Carpiodes carpio. Evermann, 1893b: 77 (Belle Fourche R. [Belle Fourche]). Evermann and Cox, 1896: 389 (characters; Emanuel Cr., Springfield). Churchill and Over, 1933: 34–35, fig. 21 (description, ecology, importance; James and larger rivers east of Missouri R. and in larger western tributaries). Shields, 1958a: 31–32; 1958b: 360 (Fort Randall Res.).

Carpiodes carpio carpio. Cleary, 1956: 286 (Big Sioux R., Lyon [Lincoln], Sioux [Lincoln and Union], and Woodbury [Union] cos., Iowa [and S. D.]). Underhill, 1959: 100 (Vermillion R.).

Ictiobus bubalus (misidentification). Evermann and Cox, 1896: 389 (characters; Crow Cr.*, [near] Chamberlain).

Carpiodes carpio is closely related to *C. velifer* (Rafinesque), and small specimens of these species are distinguished with difficulty if at all. In Iowa, Cleary (1956: 287) mapped *velifer* only from the Upper Mississippi basin except for four records from the Missouri and Big Sioux river valleys. The latter rest on probable misidentifications of young specimens of *C. carpio* made by Bailey about 1940, before he was familiar with the complexities in the taxonomy of *Carpiodes*. We know of no firm basis for the occurrence of *Carpiodes velifer* in the middle or upper parts of the Missouri basin.

Carpiodes carpio is the more abundant and widespread of the two carpsuckers in South Dakota, probably occurring in all major drainages of the Missouri system although not yet taken from the Niobrara and White basins in the state. Eddy and Surber (1947) recorded its presence in the Minnesota River, but it has not yet been found in that drainage in South Dakota and is not known from the Red River basin.

Station records: 35, 46, 51, 61, 64, 65, 72, 76, 83, 104, 120, 128, 129, 130, 132, 135, 136.

55. *Hypentelium nigricans* (LeSueur)—Northern hog sucker

Hypentelium nigricans. Churchill and Over, 1933: 28, fig. 13 (description, ecology, importance; in creeks flowing into Minnesota rivers).

Hypentelium nigricans occurs in South Dakota only in the Minnesota

River drainage. The hog sucker does not live in the Red River drainage (Underhill, 1957: 11), and it enters the Missouri River drainage only in Missouri. We have not taken specimens in South Dakota, but Dr. Underhill informs us that he collected it in the Yellowbank River near Milbank and he has loaned us for study an individual taken in the same stream in Minnesota.

KEY TO SPECIES OF *Moxostoma*

1a.—Caudal peduncle scale rows 16. [Caudal fin bright red in life]. *(Hypothetical in South Dakota.)* .... Greater redhorse, *Moxostoma valenciennesi*
1b.—Caudal peduncle scale rows typically 12. 2
  2a.—Caudal fin olive or slate-colored. Mouth moderate to large, lower lips meeting at an obtuse or sharp angle. Head moderate to large, 3.7 to 4.7 (3.3 to 3.7 in young from 1 to 3 inches) in standard length. Body scales without dark spots at base. Dorsal fin ordinarily rounded in front. 3
    3a.—Plicae of lips not broken by transverse creases into papillalike elements. Dorsal rays 11 to 15, usually 13. Dorsal base less than distance from dorsal to occiput. Body of adults yellowish. .... Golden redhorse, *Moxostoma erythrurum*
    3b.—Plicae of lips broken by transverse creases into papillalike elements. Dorsal rays 14 to 17, usually 15 or 16. Dorsal base about equal to distance from dorsal to occiput. Body of adults silvery. *(Hypothetical in South Dakota.)* .... Silver redhorse, *Moxostoma anisurum*
  2b.—Caudal fin bright red in life. Mouth small, the plicate lower lips meeting in a straight line posteriorly. Head small and subconical, 4.3 to 5.4 (3.5 to 3.8 in young from 1 to 3 inches long) in standard length. Body scales on upperparts each with a dark spot at base. Dorsal fin falcate and pointed in front. [Dorsal rays 12 to 14, usually 13]. ...... Northern redhorse, *Moxostoma macrolepidotum macrolepidotum*

56. *Moxostoma erythrurum* (Rafinesque)—Golden redhorse

Moxostoma aureolum (misidentification). Churchill and Over, 1933: 33, figs. 18–19 [in part] (description, ecology, importance; streams east of Missouri R.).

Moxostoma erythrurum has been taken by us only in the Minnesota River drainage; but since Cleary (1956: 287) found it in the Rock River, Iowa, a tributary to the Big Sioux River, the species may be expected to occur in that drainage in South Dakota.

Station records: 7, 8.

57. *Moxostoma macrolepidotum macrolepidotum* (LeSueur)— Northern redhorse

Ptychostomus haydeni. Girard, 1856: 8–9 (original description; [Missouri R.] Ft. Pierre, "Nebraska"). Girard, 1858: 220–22, pl. 49, figs. 1–4 (description; [Missouri R.] Ft. Pierre, "Nebraska").
Moxostoma duquesnei (misidentifications). Meek, 1892: 246 (Big Sioux R., Sioux Falls, S. D., and Sioux City, Iowa [and S. D.]).

Moxostoma macrolepidotum duquesnei (misidentifications). Evermann, 1893b: 77 (Belle Fourche R. and Redwater R. [Belle Fourche]; "S. Fork of" Cheyenne R. [Cheyenne Falls]). Woolman, 1896: 351 (Little Minnesota R., Browns Valley, Minn. [and S. D.] and Big Stone L., Creagers farm and Ortonville, Minn. [and S. D.]).

Moxostoma aureolum (misidentifications). Evermann and Cox, 1896: 394–95 (characters; James R., Mitchell; Emanuel and Choteau crs., Springfield; Crow Cr., [near] Chamberlain; Belle Fourche R.* and Redwater R., Belle Fourche; Cheyenne R., Cheyenne Falls). Churchill and Over, 1933: 33, figs. 18–19 [in part] (description, ecology, importance; streams east of Missouri R.).

Moxostoma aureolum aureolum (misidentifications). Cleary, 1956: 288 (Big Sioux R., Lyon [Lincoln], Sioux [Lincoln and Union], Plymouth [Union], and Woodbury [Union] cos., Iowa [and S. D.]). Underhill, 1959: 100 (Vermillion R., from Parker to mouth).

Moxostoma breviceps (misidentification). Churchill and Over, 1933: 33, fig. 20 (description; streams east of Missouri R.).

The northern redhorse, long known as *Moxostoma aureolum*, is the common species of the genus in South Dakota, and probably occurs in all drainages. We have no record from Lake Traverse, but it lives in the Red River. This redhorse is apparently not well suited to life in the Missouri River proper, but occasional specimens have been taken by fishery biologists studying Fort Randall and Oahe reservoirs.

The trinomial is adopted in recognition of *Moxostoma macrolepidotum pisolabrum* (Trautman and Martin, 1951; Minckley and Cross, 1960).

Station records: 7, 8, 46, 61, 94, 117, 118, 120, 128, 129, 132, 135, 136.

KEY TO THE SPECIES OF *Catostomus*

1a.—Scales large, in 55 to 65 rows along body and in 16 to 20 rows around caudal peduncle. Snout scarcely projects beyond upper lip. Dorsal rays 11 or 12, rarely 10. Peritoneum silvery to dusky. .... White sucker, *Catostomus commersoni*

1b.—Scales small, in 95 to 110 rows along body and in 25 to 29 rows around caudal peduncle. Snout projects notably beyond upper lip. Dorsal rays typically 10, occasionally 11. Peritoneum black. .... Longnose sucker, *Catostomus catostomus*

58. *Catostomus commersoni* (Lacépède)—White sucker

Catostomus teres. Meek, 1892: 246 (Big Sioux R., Sioux City, Iowa [and S. D.]). Woolman, 1896: 351 (Little Minnesota R.*, near [Sisseton] Indian Agency, S. D., and Browns Valley, Minn. [and S. D.]; Big Stone L., Creagers farm and Ortonville, Minn. [and S. D.]).

Catostomus teres sucklii. Evermann, 1893b: 77 (Middle Cr. and Belle Fourche R. [Belle Fourche]; Crow and Chicken crs. [Spearfish]; Rapid Cr. [Rapid City]; Cottonwood Cr. [Edgemont]; Hat Cr. [Ardmore]).

Catostomus commersonnii sucklii. Hubbs, Hubbs, and Johnson, 1943: 37–39 (hybridization; characters; stations 106 and 108 of this paper).

Catostomus commersonii. Evermann and Cox, 1896: 392–94 (characters, variation; James R., Enemy, Firesteel, and Rock crs., Mitchell; Crow Cr. [near] Chamberlain; Prairie

Cr.*, Scotland; Emanuel and Choteau crs., Springfield; French Cr.*, Custer; [Spring] Cr., Hill City; Beaver Cr., Buffalo Gap; Cheyenne R., Hot Springs and Edgemont; Redwater R. and Crow Cr., Spearfish). Churchill and Over, 1933: 28–29, fig. 14 (description, ecology, importance; most of the waters of the state). Cleary, 1956: 289 (Big Sioux R., Lyon [Lincoln], Sioux [Lincoln and Union], Plymouth [Union], and Woodbury [Union] cos., Iowa [and S. D.], [as *C. commersoni*]). Shields, 1958a: 31–32 (Fort Randall Res. [as *C. commersonnii*]). Allum and Hugghins, 1959: 34 (Brant L., Lake Co. [as *C. commersoni*]). Hugghins, 1959: 21 (Big Stone L., Waubay L., Willow L., L. Poinsett, L. Goldsmith, L. Hendricks, L. Louise, Cottonwood L., Big Sioux R., College [= Six-mile] Cr., Rapid Cr. [as *C. commersoni*]). Moyle and Clothier, 1959: 178 (Lake Traverse [as *C. commersoni*]). Underhill, 1959: 100 (Vermillion R. [as *Catastomus commersoni*]).

TABLE 7

FREQUENCY DISTRIBUTION OF PRINCIPAL DORSAL-RAY COUNTS IN *Catostomus commersoni* FROM SOUTH DAKOTA

Locality	Station No.	Dorsal Rays				N	M
		10	11	12	13		
Minnesota R. Dr.:							
Big Stone L.	3, 4	1	8	14	2	25	11.7
Whetstone Cr.	7, 8	3	3	12	6	24	11.9
Yellowbank R.	9	..	7	14	2	23	11.8
Big Sioux R. Dr.:							
Six-mile Cr.	25	..	17	29	2	48	11.7
Cheyenne R. Dr.:							
Mirror L.	121	2	26	5	..	33	11.1
Redwater Cr.	123	3	19	15	1	38	11.4
N Tribs., Mo. R.:							
S Fk. Grand R.	132	9	31	9	1	50	11.0
Little Missouri R. Dr.:							
Little Missouri R.	136	8	30	4	..	42	10.9
Box Elder Cr.	137	2	17	..	..	19	10.9

Some authors have treated the white sucker as including two widespread subspecies, a Plains form, *C. c. suckleyi,* and the eastern *C. c. commersoni.* These are reported to differ in scale size (Evermann and Cox, 1896: 392–94) and in dorsal-ray count. In both characters there is considerable local and regional variation and broad overlap. In the absence of a thorough analysis of geographic variation we can see no value in separating subspecies. The reality of a geographic gradient in dorsal-ray count is indicated by Table 7, in which western populations are seen to have fewer rays. A possible modifying role of the environment is suggested by comparison of stocks at stations 121 and 123 which lie in the same small creek system and are separated by only about a mile. Mirror Lake is a cold, spring-fed pond and in it dorsal count is lower than in Redwater Creek. Although originally drawn from the same gene pool these populations are now denied free intercommunication by a small wooden dam on the outlet of Mirror Lake. The difference (0.29 ray) is significant at the T .05 level (t = 2.109 with 69 d.f.).

Catostomus commersoni is of widespread occurrence in South Dakota, having been taken by us or reported from all principal drainages. The species is uncommon in the Missouri River.

Station records: 3, 4, 7, 8, 9, 18, 24, 25B, 26A, 26C, 27, 29, 31, 37, 38, 39, 42, 46, 49, 53, 55, 56, 58, 59, 60, 61, 62, 76, 84, 86, 88, 92, 94, 98, 100, 104, 105, 106, 108, 109, 111, 114, 115, 116, 120, 121, 123, 127, 128, 129, 131, 132, 133, 136, 137.

59. *Catostomus catostomus* (Forster)—Longnose sucker

Not previously reported to occur in the Dakotas, *Catostomus catostomus* is apparently restricted in distribution in these states to a small area, well supplied with cool, spring-fed creeks, lying to the north of the Black Hills in the Belle Fourche River drainage of the Cheyenne River system. We have three collections comprising 19 specimens. Principal dorsal rays number 10 in 13 and 11 in 3. Scale counts of 9 specimens range from 96 to 109, mean 102.3, along lateral line, and from 26 to 29, mean 27.22, around caudal peduncle.

This apparently isolated population of the longnose sucker poses a problem of dispersal. Elsewhere in the region it occurs in the upper Platte River system and in the upper Missouri and the Yellowstone rivers and their major tributaries east to the Powder River in Wyoming. The species does not seem to occur in the Little Missouri drainage (Personius and Eddy, 1955). It is apparent that the Belle Fourche population is a relict that survives from a broader distribution of the species during the Pleistocene or post-Pleistocene. Climatic and ecological changes have since resulted in restriction of range in this area.

There are two plausible channels of access to the Belle Fourche River. The longnose sucker may formerly have ranged east in the Missouri River at least to the mouth of the Cheyenne, from which it ascended to the Belle Fourche by water connectives that still persist. The second route, involving stream capture of the upper part of the Little Missouri River, calls for a less extensive former distribution, one that included that stream, but did not require use of the additional downstream waterways. It is discussed elsewhere (pp. 116–17).

Station records: 123, 126, 128.

60. *Pantosteus platyrhynchus* (Cope)—Mountain sucker

Pantosteus jordani. Evermann, 1893a: 51–56, 1 fig. (original description, comparison; Whitewood Cr., Deadwood; Spearfish*, Chicken, and Crow crs., Spearfish; Belle Fourche R., Belle Fourche; Rapid Cr.*, Rapid [City]; Hat Cr., Ardmore). Evermann, 1893b: 77 (Whitewood Cr. [Deadwood]; Spearfish, Chicken, and Crow crs. [Spearfish]; Belle Fourche R. [Belle Fourche]; Rapid Cr. [Rapid City]; Hat Cr. [Ardmore]). Evermann and Cox, 1896: 389–90 (characters, variation; Whitewood Cr., Deadwood; Spearfish*, Chicken, and Crow crs., Spearfish; Belle Fourche R., Belle Fourche; Rapid Cr.*, Rapid City; Hat Cr., Ardmore; Cheyenne R., Edgemont and Hot Springs*; Beaver Cr., Buffalo Gap; Redwater R., Spearfish; [Spring] Cr., Hill City; French Cr.*, Custer). Churchill and Over, 1933: 29–30, fig. 15 (description, ecology, habitat; streams of the Black Hills). Hubbs, Hubbs, and Johnson, 1943: 37–39 (hybridization; characters; stations 106 and 108 of this paper). Hugghins, 1959: 21 (Rapid Cr.).

As indicated in the synonymy above, the only species of *Pantosteus* in the Missouri drainage has generally been known as *Pantosteus jordani*. This form, however, appears to be conspecific with *Pantosteus platyrhynchus* from the Great Basin and Columbia drainage. Whether the eastern stock is worthy of subspecific separation is under study by Mr. Gerald R. Smith.

The only records of occurrence in South Dakota are in or near the Black Hills in the Cheyenne River system, where the species is common. In his survey of Nebraska fishes, Dr. Raymond E. Johnson collected the mountain sucker only in Hat Creek, Sioux County, in the Cheyenne drainage. However, Evermann and Cox (1896: 390) reported it from Chadron Creek at Chadron, Nebraska, in the White River system. There is a specimen (USNM 76037) 105 mm. in standard length labeled Niobrara River at Marsland, Nebraska, 1893, Evermann. It was not reported by Evermann and Cox, and the provenance of the specimen is doubtful.

Station records: 103, 106, 107, 108, 109, 112, 113, 114, 115, 116, 123, 125, 126.

ICTALURIDAE—FRESHWATER CATFISHES

KEY TO GENERA OF ICTALURIDAE

1a.—Jaws equal or the upper protruding; mouth of moderate width. Pectoral spine various; never as in 1b. Preoperculomandibular canals separate, the pores 10 or 11. Anterior nasal pore located mediad to anterior nostril. Premaxillary tooth band usually a transverse bar (except in *Noturus flavus*)................................. 2

   2a.—Adipose fin with posterior margin free, not fused or continuous with caudal fin. Gill rakers 11 or more. .... *Ictalurus*

   2b.—Adipose fin a low, keellike fleshy ridge which is fused or continuous with caudal fin. Gill rakers 3 to 10. .... *Noturus*

1b.—Lower jaw projecting; the head markedly depressed and the mouth very wide. Pectoral spine strong, almost straight, anterior and posterior edges equally armed with well-developed serrae. Preoperculomandibular canals joined in a median pore on chin, the pores 12 on each side (including median pore). Anterior nasal pore located at edge of lip, well in front of anterior nostril. Premaxillary tooth band with a broad backward projection on each side, the posterior border smoothly curved. [Adipose fin large, free from caudal.]. .... *Pylodictis*

KEY TO SPECIES OF *Ictalurus*

1a.—Caudal fin more or less truncate or rounded behind, not deeply forked. Anal rays (including all anterior rudiments) 17 to 27. Jaws nearly equal. Supraoccipital bone produced backward, but failing to join anterior process from dorsal fin. 2
  2a.—Anal rays 17 to 24, usually 22 or fewer. Chin barbels dusky. Caudal fin slightly emarginate. 3
    3a.—Pectoral spine smooth or only weakly roughened posteriorly. Outer two-thirds of interradial membranes of anal fin uniformly pigmented, always darker than the rays, the fin not mottled, barred, or uniformly pigmented on both membranes and rays. Adults with the belly yellow. .... Black bullhead, *Ictalurus melas*
    3b.—Pectoral spine with rather strong posterior serrations. Black pigment on anal fin typically densest on the membranes near their margin, or in spots that form an obscure longitudinal bar near base of fin, or in faint mottlings on both rays and membranes (in pale and unmottled specimens membranes and rays are about equally pigmented). Adults with the belly white. .... Brown bullhead, *Ictalurus nebulosus*
  2b.—Anal rays 24 to 27, usually 25 or 26. Chin barbels white, rarely faintly dusky. Caudal fin rounded behind. [Black pigment on anal fin usually most pronounced in a narrower, marginal edging and in a wider bar just distal to base of fin. Fin neither mottled nor with dark dashes on interradial membranes.] (*Hypothetical in South Dakota.*). .... Yellow bullhead, *Ictalurus natalis*
1b.—Caudal fin deeply forked. Anal rays 24 to 35. Upper jaw decidedly longer than lower. Supraoccipital bone produced backward to join with anterior process from dorsal fin. 4
  4a.—Anal shorter, its base about 3.4 to 4.3 in standard length, with 24 to 32 rays. Body silvery, the young immaculate, older fish more or less heavily spotted with dark (spots often obscure in adults, especially during the breeding season). Swimbladder with 2 chambers. .... Channel catfish, *Ictalurus punctatus*
  4b.—Anal longer, its base about 2.9 to 3.1 in standard length, with 30 to 35 rays. Body silvery, nearly or quite immaculate. Swimbladder with 3 chambers. .... Blue catfish, *Ictalurus furcatus*

61. *Ictalurus melas* (Rafinesque)—Black bullhead

Ameiurus melas. Evermann and Cox, 1896: 387 (habitat; Rock and Enemy crs., Mitchell*; Firesteel [*sic*], Choteau, and Emanuel crs., Springfield; Crow Cr., [near] Chamberlain; Prairie Cr., Scotland). Churchill and Over, 1933: 58, fig. 45 (description, ecology, importance; streams and lakes east of Missouri R. and sparingly in its western tributaries). Shields, 1958a: 31; 1958b: 360 (Fort Randall Res.).

Ictalurus melas. Cleary, 1956: 305 (Big Sioux R., Lyon [Lincoln] Co., Iowa [and S. D.]). Allum and Hugghins, 1959: 34 (Brant L., Lake Co.). Hugghins, 1959: 22 (Big Stone L., Willow L., L. Norden, Oakwood L., L. Goldsmith, L. Hendricks, L. Madison, Brant L., Wall L., Beaver L., Menno L., L. Andes, Fish L., Fraiser L., Crow L., L. Louise, Cottonwood L., Mina L.; Big Sioux R.; Angostura Res.; Black Hills streams).

Moyle and Clothier, 1959: 178 (Lake Traverse). Underhill, 1959: 101 (throughout Vermillion R. drainage).

The black bullhead is one of the most abundant and generally distributed fishes in South Dakota; we have examined specimens from all principal drainages.

Station records: 1, 2, 4, 5, 8, 9, 10, 12, 15, 16, 17, 18, 23, 24, 26A, 26B, 26C, 31, 33, 34, 35, 36, 37, 38, 46, 51, 53, 55, 56, 57, 58, 61, 72, 84, 85, 86, 88, 93, 95, 97, 100, 101, 120, 127, 129, 130, 132, 136.

62. *Ictalurus nebulosus* (LeSueur)—Brown bullhead

Ameiurus nebulosus. Woolman, 1896: 351 (Little Minnesota R., Browns Valley, Minn. [and S. D.]; Big Stone L. at Creagers farm and Ortonville, Minn. [and S. D.]).
Ictalurus nebulosus. Moyle and Clothier, 1959: 178 (Lake Traverse).

We have not examined or collected specimens and know of the occurrence of the brown bullhead in South Dakota only from the above reports.

63. *Ictalurus punctatus* (Rafinesque)—Channel catfish

Pimelodus olivaceus. Girard, 1858: 211–12, pl. 41, figs. 1–3; pl. 42 (original description; [Missouri R.] Ft. Pierre, "Nebraska").
Ichthaelurus punctatus. Cope, 1879: 440 (Missouri R. pools, near mouth of Battle Cr. [? = Blue Blanket Cr.]). Fowler, 1915: 206 (Battle Cr. of the upper Missouri).
Ictalurus punctatus. Meek, 1892: 245–46 (Missouri R. and Big Sioux R., Sioux City, Iowa [and S. D.]). Evermann, 1893b: 77 (Middle Cr., Belle Fourche). Evermann and Cox, 1896: 386 (White R., [near] Chamberlain, and Choteau Cr., Springfield). Churchill and Over, 1933: 57, fig. 44 (description, ecology, importance; Missouri, White, Cheyenne, Belle Fourche, James and Big Sioux rivers especially). Cleary, 1956: 307 (Big Sioux R., Lyon [Lincoln], Sioux [Lincoln and Union], and Woodbury [Union] cos., Iowa [and S. D.]). Moyle and Clothier, 1959: 178 (Lake Traverse). Underhill, 1959: 101 (Vermillion R., from Parker to mouth).
Ictalurus lacustris (misidentifications). Shields, 1958b: 360 (Ft. Randall Res.). Hugghins, 1959: 22 (Brant L., Menno L., Ft. Randall Res.).

The channel catfish is widely distributed, especially in streams; it has been reported or we have examined specimens from all major drainages except the Minnesota River, where it has doubtless been overlooked.

Station records: 51, 64, 65, 69, 72, 73, 76, 83, 87, 88, 120, 128, 129, 130, 135, 136.

64. *Ictalurus furcatus* (LeSueur)—Blue catfish

Ictalurus furcatus. Churchill and Over, 1933: 56–57, fig. 43 (description, ecology, importance; Missouri, White, James, Big Sioux, and Cheyenne rivers).

Despite the indication of rather wide distribution suggested by the statement of Churchill and Over, *Ictalurus furcatus* is probably restricted in South Dakota to the Missouri River and perhaps the lowermost sections of larger tributaries. Anglers not infrequently confuse the unspotted, breeding adults of channel catfish with the blue catfish.

In six years of intensive netting and creel census in Fort Randall Reservoir and tailwaters, Mr. J. T. Shields saw no blue catfish. However, they have been netted in Gavins Point Reservoir.

We have taken a single small specimen, from the Missouri River at Yankton.

Station record: 76.

KEY TO SPECIES OF *Noturus*

1a.—Head moderately deep. Jaws about equal. Vertical fins not light margined. Pectoral spine with deep, long grooves that extend nearly to base. Caudal rays long, the fin broadly rounded. Pectoral soft rays usually 6 to 8; pelvic rays usually 8. Premaxillary tooth band a transverse bar. .... Tadpole madtom, *Noturus gyrinus*

1b.—Head notably depressed. Lower jaw included. Vertical fins with light margins. Pectoral spine with shallow grooves in distal half only. Caudal rays shorter, the fin more or less truncate. Pectoral soft rays usually 9 or 10; pelvic rays usually 9 or 10. Premaxillary tooth band with a long, narrow backward projection on each side; the posterior border trapezoidal. .... Stonecat, *Noturus flavus*

65. *Noturus gyrinus* (Mitchill)—Tadpole madtom

Schilbeodes gyrinus. Evermann and Cox, 1896: 388 (description; Choteau Cr., Springfield; Prairie Cr., Scotland; Enemy, Firesteel, and Rock crs., Mitchell). Churchill and Over, 1933: 60, fig. 48 (description, ecology, importance; east of the Missouri R. in small streams or quieter portions of larger ones).

Noturus gyrinus. Meek, 1892: 246 (characters; Big Sioux R. [Sioux Falls or Sioux City]).

The tadpole madtom has been taken in our survey from the Minnesota, Big Sioux, and James river drainages. It is especially partial to clear, weedy waters, a habitat now less common than formerly, and the species is now uncommon.

Station records: 8, 15, 26C, 39, 44.

66. *Noturus flavus* Rafinesque—Stonecat

Noturus flavus. Meek, 1892: 245–46 (characters; Missouri R., Sioux City, Iowa [and S. D.]; Big Sioux R., Sioux Falls). Evermann, 1893b: 77 ("S. Fork of" Cheyenne R. at Cheyenne Falls; Belle Fourche R. at Belle Fourche). Evermann and Cox, 1896: 388 (Emanuel Cr., Springfield; Cheyenne R., Cheyenne Falls; Belle Fourche R., Belle Fourche; Beaver Cr., Buffalo Gap). Churchill and Over. 1933: 56–60, fig. 47 (description, ecology, importance; larger streams of the state). Underhill, 1959: 101 (lower six miles of Vermillion R.).

The stonecat probably occurs in all major drainages except that of Lake

Traverse. It apparently reached the Big Stone Lake area too late to utilize the Warren River waterway into the Red River.

Station records: 7, 8, 56, 64, 65, 69, 72, 79, 86, 94, 120, 129, 130, 132, 135, 136, 137.

67. *Pylodictis olivaris* (Rafinesque)—Flathead catfish

Leptops olivaris. Evermann and Cox, 1896: 387 (characters; mouth of White R. [near] Chamberlain).

Opladelus olivaris. Churchill and Over, 1933: 59, fig. 46 (description, ecology, importance, compiled; [White R. near Chamberlain]).

Pylodictis olivaris. Cleary, 1956: 309 (Big Sioux R., Plymouth [Union] Co., Iowa [and S. D.]). Underhill, 1959: 101 (as *P. olivarus*; mouth of Vermillion R.).

We have not examined specimens, but the above records confirm the occasional occurrence of the flathead catfish in the Missouri River and in the lower courses of some of its larger tributaries. In 1959 Mr. Ned E. Fogle netted a 22-inch specimen in Oahe Reservoir, apparently the known upstream limit of distribution. It is reported that commercial fishermen take many large flathead catfish in large-mesh gill nets in Fort Randall Reservoir.

ANGUILLIDAE—FRESHWATER EELS
68. *Anguilla rostrata* (LeSueur)—American eel

Anguilla rostrata. Churchill and Over, 1933: 26–27, fig. 12 (description, ecology, importance; Vermillion R., Split Rock Cr., and Big Stone L.). Underhill, 1959: 101 (lower Vermillion R.).

The American eel apparently ascends the Missouri and Minnesota rivers into South Dakota only in small numbers. In 1952, state Game Warden C. E. Gunderson told us that "green eels" are caught occasionally at Yankton. We have heard anglers' reports of them from as far upstream as the mouth of the White River, but eels can no longer ascend above Gavins Point Dam. We have not seen specimens.

CYPRINODONTIDAE—KILLIFISHES
KEY TO SPECIES OF *Fundulus*

1a.—Dorsal origin ahead of vertical through anal origin, its distance from caudal base 1.3 to 1.7 in predorsal length. Dorsal rays 11 to 16. Scale rows on body 38 to 68. Body with vertical dark bars. 2

2a.—Peritoneum silvery. Intestine with a single loop, about one-half body length. Scales large, in 38 to 48 transverse series and in 27 to 33 rows around body (in advance of dorsal and pelvic fins). Postorbital canal (extending upward from behind eye)

continuous, with 4 pores. Eye larger, the orbit contained 1.3 to 1.7 in postorbital length of head. .... Banded killifish, *Fundulus diaphanus menona*
2b.—Peritoneum black. Intestine convoluted, about 1.5 times body length. Scales small, in 61 to 68 transverse series and in 53 to 65 rows around body. Postorbital canal interrupted, consisting of 1 or 2 short tubes each with 2 pores. Eye smaller, the orbit contained 2.0 to 2.5 times in postorbital length of head ....
.... Plains killifish, *Fundulus kansae*
1b.—Dorsal origin behind anal origin, its distance from caudal base 2.0 to 2.5 in predorsal length. Dorsal rays 8 to 10. Scale rows on body 34 to 36. Body without vertical bars.
.... Plains topminnow, *Fundulus sciadicus*

69. *Fundulus diaphanus menona* Jordan and Copeland—Banded killifish

Fundulus zebrinus (misidentification). Meek, 1892: 246 (Big Sioux R., Sioux City, Iowa [and S. D.]).

Fundulus diaphanus. Woolman, 1896: 352, 357 (habitat; Big Stone L., Creagers farm and Ortonville, Minn. [and S. D.]; Wheatstone [Whetstone] Cr., Milbank).

Zygonectes diaphanus menona. Churchill and Over, 1933: 63, fig. 51 (description; small sluggish streams).

The banded killifish occurs in South Dakota in lakes of the Minnesota River drainage, and on the basis of Meek's report, formerly lived in the Big Sioux drainage, to which it probably gained access from the Minnesota or Des Moines basin. Elsewhere the species is known in South Dakota only in Lake Andes (station 57). This population presumably stems from an inadvertent introduction, likely mixed with largemouth bass from a federal fish hatchery.

Station records: 11, 57.

70. *Fundulus kansae* Garman—Plains killifish

Fundulus kansae. Miller, 1955: 11–12 (introduction; Cheyenne R.*, above U. S. 14–16 bridge, Pennington Co., and at Oral*, Fall River Co.).

The status of this killifish in South Dakota has been discussed by Miller (*op. cit.*), who concluded that it is probably introduced. It is known only in the Cheyenne River system.

Station records: 104, 120.

71. *Fundulus sciadicus* Cope—Plains topminnow

Fundulus sciadicus. Evermann and Cox, 1896: 416 (description, habitat; Prairie Cr.*, Scotland; Rock Cr., Mitchell).

Zygonectes sciadicus. Churchill and Over, 1933: 62–63, fig. 50 (description, ecology, importance; smaller streams east of Missouri R.).

Fundulus sciadicus was formerly rather common in clear creeks in southern South Dakota, but appears now to be uncommon.

Station records: 35, 52, 84, 85.

GADIDAE—CODFISHES
72. *Lota lota* (Linnaeus)—Burbot

Lota maculosa. Cope, 1879: 440 (Battle [? = Blue Blanket] Cr.). Churchill and Over, 1933: 80, fig. 71 (description, ecology, importance; mouth of Rapid Cr. and Missouri R.).

Lota lota maculosa. Evermann, 1893b: 78 (Cheyenne Falls [Cheyenne R.]). Evermann and Cox, 1896: 424 (Cheyenne R. at Cheyenne Falls).

Lota lota. Shields, 1958a: 31–32 (Fort Randall Res.). Underhill, 1959: 101 (mouth of Vermillion R.).

Burbot were formerly taken below the Black Hills in the Cheyenne River system, but we know of no recent records except in or adjacent to the Missouri River, where the species is common.

Station records: 52, 64.

GASTEROSTEIDAE—STICKLEBACKS
73. *Culaea inconstans* (Kirtland)—Brook stickleback

Eucalia inconstans. Evermann and Cox, 1896: 416 (description; Crow Cr., [near] Chamberlain). Woolman, 1896: 348, 352 (coloration; Daugherty Cr., Browns Valley, Minn. [and S. D.]; Big Stone L., Ortonville, Minn. [and S. D.]). Churchill and Over, 1933: 63–64, fig. 52 (description, ecology, importance; streams and lakes east of Missouri R. and a few small streams west of that river). Cleary, 1956: 160, 324 (description, ecology, importance; Big Sioux R., Lyon [Lincoln] and Sioux [Lincoln and Union] cos., Iowa [and S. D.]). Underhill, 1959: 102 (West Fork of Vermillion R.).

It has been shown by Whitley (1950: 44) that *Eucalia* Jordan, 1878, in the Gasterosteidae is preoccupied by *Eucalia* C. Felder, 1861, in Lepidoptera (Nymphalidae). Barring a redefinition of generic limits in the Gasterosteidae we see no suitable alternative to the adoption of *Culaea* Whitley, 1950, as the generic name for the brook stickleback.

Culaea inconstans typically lives in rather clear, sluggish or standing, spring-fed waters that are well vegetated, a diminishing habitat in South Dakota. This ecological restriction rather than zoogeographic limitation probably explains the absence of firm records in western South Dakota. The species lives in northern Nebraska (Johnson, MS), is widely distributed in North Dakota (Hankinson, 1929: 456), and occurs in eastern Montana. It survives as a scattered relict in suitable persistent waters on the Great Plains.

Station records: 3, 8, 10, 19, 23, 25B, 28, 35.

PERCOPSIDAE—TROUT-PERCHES

74. *Percopsis omiscomaycus* (Walbaum) —Trout-perch

Percopsis guttatus. Meek, 1892: 246 (Big Sioux R.*, Sioux City, Iowa [and S. D.]). Woolman, 1896: 352 (Little Minnesota R., Browns Valley, Minn. [and S. D.]; Big Stone L., Creagers farm, Minn. [and S. D.]).

Percopsis omiscomaycus. Churchill and Over, 1933: 64–65, fig. 53 (description, ecology, importance; northeastern S. D.). Cleary, 1956: 158, 311 (description; Big Sioux R., Sioux [Lincoln and Union] and Plymouth [Union] cos., Iowa [and S. D.]).

The trout-perch has been collected in South Dakota only in the Minnesota and Big Sioux river drainages. We have re-examined specimens reported by Woolman from the Minnesota River at Ortonville, Minnesota (UMMZ 177414). The species apparently gained access to the Missouri drainage in South Dakota and in northwestern Iowa by crossover connection from the Minnesota River drainage; there are no records from the Des Moines basin (Cleary, 1956: 311), and Johnson (MS) did not take *Percopsis omiscomaycus* in Nebraska. Records from southern Iowa, eastern Kansas, and northern Missouri indicate that the species entered the lower part of the Missouri basin in an independent invasion.

Station records: 15, 16.

SERRANIDAE—BASSES

75. *Roccus chrysops* (Rafinesque)—White bass

Roccus chrysops. Meek, 1892: 246 (Big Sioux R., Sioux City, Iowa [and S. D.]). Cox, 1896: 611 (Big Stone L.). Woolman, 1896: 350 (Big Stone L. and Little Minnesota R.).

Lepibema chrysops. Churchill and Over, 1933: 73–74, fig. 63 (description, ecology, importance; lakes and larger streams of eastern S. D.). Eddy and Surber, 1947: 201–02 (description, habits; Big Stone L.).

Morone chrysops. Hugghins, 1959: 22 (Big Stone L.).

The white bass is limited in original distribution in South Dakota to the Minnesota and Big Sioux river drainages, where it is largely restricted to lakes. It probably crossed into the middle-Missouri system by stream crossover through the lake area of southwestern Minnesota (see pp. 122–25).

Transplanting has greatly enlarged the natural range. Introductions have been made into many eastern lakes, the Missouri River impoundments, and some western waters, including Shadehill Reservoir.

Station records: 3, 8, 15, 16.

CENTRARCHIDAE—SUNFISHES

KEY TO GENERA OF CENTRARCHIDAE

1a.—Anal spines 3 (rarely 2 or 4). Dorsal spines usually 10. 2
  2a.—Body elongate, depth 3 to 5 in standard length (somewhat deeper in large adults). Lateral-line scales more than 55. Precaudal vertebrae typically 15. .... *Micropterus*
  2b.—Body compressed, oblong; depth usually 2.0 to 2.5 in standard length. Lateral-line scales fewer than 55. Precaudal vertebrae typically 12. .... *Lepomis*
1b.—Anal spines 5 to 7, usually 6. Dorsal spines not 10. 3
  3a.—Dorsal spines 11 or 12; base of anal contained 1.7 to 2.0 times in base of dorsal. Gill rakers moderate in length, fewer than 15. Branchiostegal rays 6. Preopercle nearly entire. .... *Ambloplites*
  3b.—Dorsal spines 5 to 9; base of anal about equal to base of dorsal. Gill rakers long and slender, more than 30. Branchiostegal rays 7. Preopercle finely serrate. ..*Pomoxis*

KEY TO SPECIES OF *Micropterus*

1a.—Outline of spinous dorsal gently curving, the shortest spine at emargination more than half as long as the longest. Anal and soft dorsal with scales on membranes near base. Pyloric caeca typically unbranched. Scales smaller, 68 to 81 along lateral line, and 14 to 18 rows on cheek from eye to angle of preopercle. Pattern consists principally of vertical dark bars that become obscured with age; young with base of caudal yellow, succeeded by a marked dark band and the edge of fin clear white. .... Smallmouth bass, *Micropterus dolomieui*
1b.—Outline of spinous dorsal angulate, the shortest spine at emargination less than half as long as longest. Anal and soft dorsal normally without scales on membranes near base. Pyloric caeca typically branched at base. Scales larger, 58 to 69 along lateral line, and 9 to 12 rows on cheek from eye to angle of preopercle. Pattern consists chiefly of a rather regular longitudinal dark stripe on side; young without marked band on caudal. .... Largemouth bass, *Micropterus salmoides salmoides*

76. *Micropterus dolomieui* Lacépède—Smallmouth bass

Micropterus dolomieu. Woolman, 1896: 352 (Big Stone L., Creagers farm, Minn. [and S. D.]; Little Minnesota R., Browns Valley, Minn. [and S. D.]). Churchill and Over, 1933: 73, fig. 62 (description, ecology, importance, presumptive report; Big Stone L.).

Woolman's records of the smallmouth bass from the upper Minnesota River drainage might be doubted because they were based on very small specimens. We have recently re-examined part of Woolman's material from the Minnesota River, Ortonville, Minnesota, and confirm his identification of this species (UMMZ 177419) as well as that of *Micropterus salmoides* (UMMZ 177423).

The smallmouth bass is apparently confined in South Dakota to the Minnesota River drainage, and is not known to have been introduced successfully elsewhere. The only presumably native record known to us from the middle or upper Missouri basin is that of Meek (1894: 138) from Spirit

Lake, Iowa, a locality to which the smallmouth probably gained access by past stream connections with the Minnesota or Des Moines rivers.

77. *Micropterus salmoides salmoides* (Lacépède)—Largemouth bass

Micropterus salmoides. Woolman, 1896: 352 (Big Stone L., Creagers farm, Minn. [and S. D.]; Little Minnesota R., Browns Valley, Minn. [and S. D.]). Shields, 1958a: 31; 1958b: 360 (Fort Randall Res.). Hugghins, 1959: 22 (Big Stone L., L. Chapelle, Mina L., Red Plum L., Ft. Randall Res.). Moyle and Clothier, 1959: 178 (Lake Traverse).

Micropterus salmoides salmoides. Underhill, 1959: 101 (Vermillion R., from Centerville to mouth).

Huro floridana. Churchill and Over, 1933: 72, fig. 61 (description, ecology, importance; larger lakes of the state).

There is no firm basis to indicate that the largemouth bass was originally present in the Missouri basin in South Dakota. Meek (1894: 138) reported it from Spirit Lake, and from the Floyd River at LeMars and Sioux City, northwestern Iowa. It is not certain that these are natural occurrences, however, since Meek mentioned plantings of largemouth bass in Storm Lake, Iowa, about five years prior to his visit in 1890 (*op. cit.*: 134). Johnson (MS) suspected that the largemouth bass may be native in parts of Nebraska because of a collection in 1872 from near Omaha and collections listed by Evermann and Cox, 1896: 419, from the Platte River basin. As the result of active plantings of this species for many years the largemouth is now known to occur in all principal drainages in South Dakota except that of the Little Missouri River.

Station records: 1, 4, 8, 14, 36, 37, 40, 54, 57, 62, 63, 72, 84, 100, 134.

KEY TO SPECIES OF *Lepomis*

1a.—Opercle (not including membrane) stiff to its margin; not fimbriate along posterior edge. 2

    2a.—Pectoral short and broadly rounded; about 4 in standard length. Gill rakers moderately long and slender, the longest if depressed extending to base of second (third in young) raker below. Opercle broadly margined with light, without scarlet in life. Supramaxilla about two-thirds breadth of maxilla. Inferior pharyngeal bone elongate, external margin straight, teeth rather sharp. Palatine teeth fairly well developed. .... Green sunfish, *Lepomis cyanellus*

    2b.—Pectoral long and pointed; 3.0 to 3.3 in standard length. Gill rakers short and stout, the longest if depressed extending to base of first (second in young) raker below. Opercular margin dark, with a small, semicircular scarlet spot. Supramaxilla about one-third breadth of maxilla. Inferior pharyngeal bone broad and heavy, the external margin a sigmoid curve, teeth blunt. Palatine teeth normally absent (often a single tooth developed). .... Pumpkinseed, *Lepomis gibbosus*

1b.—Opercle produced into a thin, flexible projection lying within the opercular membrane; often more or less fimbriate or ragged posteriorly. 3

3a.—Gill rakers short and stout, knoblike; the longest when depressed not extending beyond first raker below (except in young). Longest anal spine usually 1.8 to 2.4 (1.4 or more in young) in distance from insertion of pelvic to origin of anal. Pectoral short, obovate. Caudal vertebrae typically 18. (*Hypothetical in South Dakota.*).
.... Longear sunfish, *Lepomis megalotis peltastes*
3b.—Gill rakers rather long and slender, the longest when depressed extending to base of second raker below (third in young). Longest anal spine usually 1.0 to 1.8 in distance from insertion of pelvic to origin of anal (1.0 to 1.4 in young). Pectoral moderate to long. Caudal vertebrae typically 17. 4
4a.—Opercle extending little into membranous flap, its margin entire; opercular membrane broadly margined with light. Anal III, 7 to 9. No dark blotch on posterior dorsal rays. Palatine teeth present. Sensory cavities of head well developed, the supraorbital canals wider than interspace. ....
.... Orangespotted sunfish, *Lepomis humilis*
4b.—Opercle extending almost to membranous margin, edge of opercle fimbriate; opercular membrane dark to its margin. Anal III, 10 to 12. A dark blotch on median part of posterior dorsal rays. Palatine teeth absent. Sensory cavities of head not enlarged, the supraorbital canals much narrower than interspace.
.... Bluegill, *Lepomis macrochirus*

78. *Lepomis cyanellus* Rafinesque—Green sunfish

Lepomis cyanellus. Meek, 1892: 246 (Big Sioux R., at Sioux Falls, S. D. and Sioux City, Iowa* [and S. D.]). Cleary, 1956: 314 (Big Sioux R., Lyon [Lincoln], Sioux [Lincoln and Union], Plymouth [Union], and Woodbury [Union] cos., Iowa [and S. D.]). Shields, 1958a: 31 (Fort Randall Res.). Underhill, 1959: 101 (Vermillion R. drainage).

Apomotis cyanellus. Evermann and Cox, 1896: 417–18 western limit of distribution, coloration; Rock, Enemy*, and Firestone* crs., and James R., Mitchell; Emanuel and Choteau crs., Springfield; Prairie Cr.*, Scotland). Churchill and Over, 1933: 69, fig. 58 (description, ecology, importance; smaller streams of eastern half of state).

The older records cited above support the claim that the green sunfish was native to the Missouri drainage basin in eastern South Dakota. Presently the species also lives in western drainages except the Little Missouri, but these occurrences probably result at least in part from intentional stocking or inadvertent plantings with other fishes. We have no records from the Minnesota or Red River basins.

Station records: 31, 33, 34, 35, 37, 38, 39, 40, 43, 49, 51, 53, 55, 58, 61, 65, 72, 83, 84, 86, 90, 93, 97, 98, 100, 120, 122, 127, 128, 129, 130, 133, 134.

79. *Lepomis gibbosus* (Linnaeus)—Pumpkinseed

Eupomotis gibbosus. Churchill and Over, 1933: 68–69, fig. 57 (description, ecology, importance; smaller streams of eastern part of state).

Lepomis gibbosus. Harrison and Speaker, 1954: 521 (Big Sioux R., Lyon [Lincoln] Co., Iowa [and S. D.]). Cleary, 1956: 315 (Big Sioux R., Lyon [Lincoln] Co., Iowa [and S. D.]).

Although the pumpkinseed was probably native to the Minnesota River drainage and Meek (1894: 137) reported it from Spirit Lake, in the Missouri basin of Iowa, there is no confirmation that it was native to the Missouri drainage in South Dakota. It is now present, through probable or known plantings, in the Big Sioux, James, White, and Grand river drainages.

Station records: 4, 8, 33, 37, 93, 95, 133.

80. *Lepomis macrochirus* Rafinesque—Bluegill

Lepomis pallidus. Woolman, 1896: 352 (Big Stone L., Ortonville, Minn. [and S. D.]).
Helioperca incisor. Churchill and Over, 1933: 70, fig. 59 (description, ecology, importance; larger lakes and rivers of eastern half of state).
Lepomis macrochirus. Cleary, 1956: 315 (Big Sioux R., Sioux [Lincoln] Co., Iowa [and S. D.]). Shields 1958a: 31; 1958b: 360 (Fort Randall Res.). Hugghins, 1959: 23 (Big Stone L., Wilmarth L., Menno L., Fish L., L. Louise, Mina L., Red Plum L., Angostura Res., Black Hills streams). Moyle and Clothier, 1959: 179 (Lake Traverse). Underhill, 1959: 101 (Vermillion R. basin).

The bluegill was native to the Minnesota River drainage and, presumably, Lake Traverse, but it is probable that occurrence in most other South Dakota waters (Table 9) results from plantings. Meek reported it (1894: 137) from Spirit Lake, Iowa. The species is now widely distributed in South Dakota.

Station records: 1, 3, 4, 5, 8, 16, 36, 37, 54, 63, 93, 134.

81. *Lepomis humilis* (Girard)—Orangespotted sunfish

Lepomis humilis. Meek, 1892: 246 (Big Sioux R., at Sioux Falls, S. D. and Sioux City*, Iowa [and S. D.]). Evermann and Cox, 1896: 418 (coloration; Rock* and Firesteel crs. and James R., Mitchell; Prairie Cr., Scotland). Cleary, 1956: 136, 316 (description, life history, importance; Big Sioux R., Lyon [Lincoln], Sioux [Lincoln and Union], and Woodbury [Union] cos., Iowa [and S. D.]). Shields, 1958a: 31 (Fort Randall Res.). Moyle and Clothier, 1959: 179 (Lake Traverse). Underhill, 1959: 101 (Vermillion R. drainage).
Allotis humilis. Churchill and Over, 1933: 71, fig. 60 (description, ecology, importance; rivers and creeks of eastern half of state [as *Allotis humulis*]).
Lepomis megalotis (misidentifications). Woolman, 1896: 352, 358 ([Big Stone L.], Ortonville; Wheatstone [Whetstone] Cr., Milbank).

Lepomis humilis is native to all principal eastern drainages of South Dakota. Evermann and Cox (1896: 418) noted that no sunfish inhabited seemingly suitable ponds near Chamberlain, and remarked that *L. humilis* did not range as far west as did *L. cyanellus*. The present widespread occurrence of the orangespotted sunfish west of the Missouri River results from inadvertent plantings with other species.

Station records: 1, 9, 14, 15, 20, 21, 22, 23, 24, 26A, 26C, 27, 30, 31, 33, 34, 35, 39, 40, 43, 46, 49, 51, 53, 58, 59, 61, 62, 72, 83, 96, 120, 133, 134.

82. *Ambloplites rupestris rupestris* (Rafinesque)—Rock bass

Ambloplites rupestris. Meek, 1892: 246 (Big Sioux R., at Sioux Falls, S. D., and Sioux City, Iowa [and S. D.]). Churchill and Over, 1933: 67–68, fig. 56 (Big Stone L. and [Minnesota R. drainage]). Hugghins, 1959: 23 (Rapid Cr.).

On the basis of Meek's records, the rock bass apparently once lived in the Big Sioux basin, which it probably entered from the Minnesota River drainage, where it is native. It has been established through stocking in Rapid Creek in the Black Hills, and lives in Pickerel and perhaps other lakes in the upper part of the Big Sioux drainage.

Station records: 7, 8.

KEY TO SPECIES OF *Pomoxis*

1a.—Dorsal spines normally 6. Dorsal base much shorter than distance from origin of dorsal to back of eye (58 to 65 per cent of predorsal length). Caudal vertebrae typically 18. Mouth moderately oblique. .... White crappie, *Pomoxis annularis*
1b.—Dorsal spines typically 7 or 8. Dorsal base equal to or greater than distance from origin of dorsal to back of eye (73 to 81 per cent of predorsal length). Caudal vertebrae typically 19. Mouth strongly oblique. .... Black crappie, *Pomoxis nigromaculatus*

83. *Pomoxis annularis* Rafinesque—White crappie

Pomoxis annularis. Woolman, 1896: 352 (Big Stone L., Creagers farm, Minn. [and S. D.]). Churchill and Over, 1933: 67, fig. 55 (description; streams and lakes east of Missouri R.). Cleary, 1956: 138, 317 (description, life history, importance; Big Sioux R., Lyon [Lincoln], Sioux [Lincoln and Union], Plymouth [Union], and Woodbury [Union] cos., Iowa [and S. D.]). Shields, 1958a: 31; 1958b: 360 (Fort Randall Res.). Hugghins, 1959: 23 (Big Stone L., L. Goldsmith, L. Chapelle, L. Louise). Moyle and Clothier, 1959: 178 (Lake Traverse). Underhill, 1959: 102 (Vermillion R. basin).

The only old record of the white crappie in South Dakota is that by Woolman from Big Stone Lake. In 1896 (p. 417) Evermann and Cox wrote: "Both species of *Pomoxis* are being extensively introduced into the waters of Kansas, Nebraska, and South Dakota, and it is not easy to determine definitely the natural western limit of either. It will be very close to the truth, however, if we put it in the eastern part of Nebraska and the Dakotas on the border of the alkali region." Both crappies are now widely distributed in South Dakota.

Station records: 1, 3, 4, 7, 8, 9, 15, 39, 43, 46, 50, 51, 61, 65, 72.

84. *Pomoxis nigromaculatus* (LeSueur)—Black crappie

Pomoxis sparoides (misidentification). Churchill and Over, 1933: 66, fig. 54 (description, ecology, importance; east of Missouri R.).
Pomoxis nigromaculatus. Shields, 1958a: 31; 1958b: 360 (Fort Randall Res.). Hugghins, 1959: 23 (Big Stone L., L. Hendricks, Brant L., Wilmarth L., L. Louise, Mina L., Fraiser L., Angostura Res.). Moyle and Clothier, 1959: 178 (Lake Traverse). Underhill, 1959: 102 (Vermillion R. basin).

In the absence of old records and in view of the early stocking of the species (see p. 99), it is not possible now to be sure whether the black crappie was native to South Dakota, but it probably lived at least in the Minnesota River drainage. Meek reported the species from East Okoboji Lake, Iowa (1894: 137).

Station records: 1, 2, 3, 7, 8, 14, 16, 37, 38, 39, 46, 50, 51, 54, 61, 72, 84, 134.

PERCIDAE—PERCHES

KEY TO GENERA OF PERCIDAE

1a.—Preopercle strongly serrate. Branchiostegal rays 7 (rarely 8). No distinct urogenital papilla. Top of skull ridged; supraoccipital crest high. Fishes of medium to large size. 2

2a.—Strong canine teeth on jaws and palatine. Pelvic fins widely separated (interspace equal to breadth of fin base). Body slender and subterete. Anal II, 12 or 13. Pseudobranchium well developed. .... *Stizostedion*

2b.—No canine teeth. Pelvic fins close together. Body rather deep and compressed, crossed by about 7 prominent vertical dark bands. Anal II, 6 to 8. Pseudobranchium rudimentary. .... *Perca*

1b.—Preopercle nearly or quite entire. Branchiostegal rays 6 (rarely 5). Genital papilla prominent. Top of head nearly or quite smooth; supraoccipital crest weak or absent. Fishes of small size, the largest only 6 or 7 inches long, most much smaller. [Pseudobranchium rudimentary or absent.] 3

3a.—Interpelvic space and belly either naked or with enlarged and modified median scales which are strongly ctenoid (modified scales sometimes little enlarged and occasionally of normal size in females, but at least one enlarged interpelvic scale typically present). Anal fin large, about equal to or larger than soft dorsal (somewhat smaller in *P. caprodes*). Lateral line complete. Vertebrae 39 to 45 (rarely 38). .... *Percina*

3b.—Breast, interpelvic space, and belly variously naked or covered with normal scales, but never with a midventral series of enlarged and modified scales. Anal fin smaller than soft dorsal. Lateral line often incomplete. Vertebrae 33 to 40 (rarely 41). .... *Etheostoma*

KEY TO SPECIES OF *Stizostedion*

1a.—Lower lobe of caudal not tipped with white. Spinous dorsal with clear-cut black spots (except in young), but without a large black blotch near base of posterior spines. Second dorsal rays 18 to 22 (usually 19 or 20). Cheek usually well scaled. Pyloric caeca

3 to 9 (usually 5), each shorter than stomach. Back with 3 or 4 dark saddles, these expanded laterally to form 3 prominent oblong blotches—one below each dorsal fin and a smaller one on caudal peduncle. .... Sauger, *Stizostedion canadense*
1b.—Lower lobe of caudal with a milk-white tip. Spinous dorsal without clearly defined black spots; a large black blotch near base of posterior spines. Second dorsal rays 20 to 22 (usually 21 or 22). Cheek usually with few scales. Pyloric caeca 3, each about as long as stomach. Back crossed with 6 or 7 narrow dark saddles. ....
.... Walleye, *Stizostedion vitreum vitreum*

85. *Stizostedion canadense* (Smith) —Sauger

Stizostedion canadense. Meek, 1892: 246 (Big Sioux R. [Sioux Falls, S. D., or Sioux City, Iowa]). Cleary, 1956: 318 (Big Sioux R., Woodbury [Union] Co., Iowa [and S. D.]). Shields, 1958a: 31; 1958b: 360 (Fort Randall Res.). Hugghins, 1959: 23 (Ft. Randall Res.). Underhill, 1959: 102 (Vermillion R., from Centerville to mouth).
Stizostedion vitreum (misidentification). Meek, 1892: 246 (Big Sioux R.*, Sioux City, Iowa [and S. D.]).
Stizostedion canadense boreum. Evermann and Cox, 1896: 419–20 (pyloric caeca count; Choteau Cr., Springfield; White R., [near] Chamberlain).
Cynoperca grisea. Churchill and Over, 1933: 76 (description, habitat, importance; western rivers and Missouri R.; presumptive report from lakes).

The sauger is a common species throughout the Missouri River in South Dakota from which it moves upstream for short distances in some tributaries. The large dams on the Missouri now block upstream migrations of the sauger and large sport fisheries for it have developed in the reservoirs and in the tailwaters. For example, below Oahe Reservoir in the period July 1959 through March 1960, 31,291 sauger that averaged over two pounds were harvested; many weighed from 4 to 7 pounds (Fogle, MS). Although the sauger occurs in the Minnesota River in Minnesota it is not known from Big Stone Lake. In South Dakota it apparently does not live in natural lakes other than ox-bows.

Station records: 64, 65, 72, 76, 135.

86. *Stizostedion vitreum vitreum* (Mitchill)—Walleye

Stizostedion vitreum. Evermann and Cox, 1896: 419 (habitat, pyloric caeca count; Crow Cr., [near] Chamberlain; Rock Cr., Mitchell; Choteau Cr., Springfield). Woolman, 1896: 353 (Big Stone L., Creagers farm, Minn. [and S.D.]). Churchill and Over, 1933: 75–76, fig. 65 (description, ecology, importance; lakes of eastern part of state). Hugghins, 1959: 23 (Big Stone L., Waubay L., L. Goldsmith, L. Hendricks, Brant L., L. Madison, Fish L., Cottonwood L., Mina L.). Moyle and Clothier, 1959: 178 (Lake Traverse).
Stizostedion vitreum vitreum. Allum and Hugghins, 1959: 34 (Brant L., Lake Co.). Underhill, 1959: 102 (Lake Marindahl [Yankton Co.]).

On the basis of early reports the walleye may be interpreted as a native inhabitant of many lakes and rivers of eastern South Dakota, including the Missouri River, where the species persists but in smaller numbers than the sauger. It is probable that the walleye enters the lower parts of western rivers, but our only record is from the Grand River. Plantings in artificial reservoirs in western South Dakota, including Angostura, Belle Fourche, and Shadehill, have been successful, sometimes spectacularly so.

Station records: 7, 9, 15, 22, 64, 65, 72, 101, 128, 135.

87. *Perca flavescens* (Mitchill)—Yellow perch

Perca flavescens. Meek, 1892: 246 (Big Sioux R., Sioux Falls, S. D., and Sioux City, Iowa [and S. D.]). Evermann and Cox, 1896: 420 (Enemy, Rock, and Firesteel* crs., and James R., Mitchell). Woolman, 1896: 353, 358 (Big Stone L., Creagers farm and Ortonville, Minn. [and S. D.]; Little Minnesota R., near [Sisseton] Indian Agency, S. D., and Browns Valley, Minn. [and S. D.]). Churchill and Over, 1933: 74–75, fig. 64 (description, ecology, importance; nearly all lakes and many streams of eastern half of state). Shields, 1958a: 31; 1958b: 360 (Ft. Randall Res.). Allum and Hugghins, 1959: 34 (Brant L., Lake Co.). Hugghins, 1959: 23 (Waubay L., Willow L., Lake Goldsmith, Crow L., Cottonwood L., Mina L., McNeil Pond, Medary Cr., Ft. Randall Res., Angostura Res.). Moyle and Clothier, 1959: 178 (Lake Traverse). Underhill, 1959: 102 (Vermillion R. basin).

On the basis of the older records the yellow perch is interpreted as native to eastern South Dakota as far west as the James River (Evermann and Cox, 1896: 420). Our only record for the Missouri River prior to impoundment (station 65) is from a backwater adjacent to a bottomland lake that was connected to the river during a spring flood the year of our collection. Johnson (MS) regarded the perch as doubtfully native to Nebraska, and the few records from Kansas and southwestern Iowa presumably result from introductions. We believe that the perch gained access to the middle Missouri basin postglacially from either the Minnesota or the Des Moines basin (see p. 125). The perch has been widely planted and now lives in most waters to which it is well adapted. In addition to the collections listed below, Allum has seen specimens from Dog Ear Lake, Tripp County; Lacreek Lake, Bennett County; Burke Lake, Gregory County; Murdo Lake, Jones County; and Presho Lake, Lyman County.

Station records: 1, 2, 3, 4, 8, 9, 11, 13, 14, 16, 17, 20, 21, 22, 23, 24, 30, 31, 33, 34, 37, 38, 39, 41, 46, 47, 58, 61, 65, 101, 128, 130, 134.

KEY TO SPECIES OF *Percina*

1a.—Interorbital space neither especially broad nor depressed. Snout not projecting beyond upper jaw. Lateral-line scales fewer than 78. 2

2a.—Belly mostly scaled and with the scales of the mid-line strongly modified (at least in adult males). Premaxillary frenum broad, not hidden by a cross furrow. Anal fin of adult male not notably elevated, without tubercles. 3

3a.—Gill membranes separate; distance from junction to tip of mandible less than that to insertion of pelvic. Snout rather blunt, more or less decurved. Spinous dorsal without contrasting orange band; a dark blotch near front of fin. ....
.... Blackside darter, *Percina maculata*
3b.—Gill membranes broadly connected; distance from junction to tip of mandible greater than that to insertion of pelvic. Snout long and sharply pointed. Spinous dorsal with an orange submarginal band; no anterior dark blotch. ....
.... Slenderhead darter, *Percina phoxocephala*
2b.—Belly largely scaleless medially, but usually crossed before anus by a bridge of scales; scales of mid-line little modified. Premaxillary frenum very narrow or hidden by a furrow behind upper lip. Anal fin of adult male excessively elevated, the tips of the longest rays reaching approximately to base of caudal fin, with prominent tubercles during the breeding season. (*Hypothetical in South Dakota.*) ....
.... River darter, *Percina shumardi*
1b.—Interorbital space broad, more or less depressed. Snout forming a conical fleshy protuberance which projects beyond upper jaw. Lateral-line scales 78 or more. ....
.... Logperch, *Percina caprodes semifasciata*

88. *Percina maculata* (Girard)—Blackside darter

Etheostoma aspro. Meek, 1892: 246 (Big Sioux R. [Sioux Falls, S. D., or Sioux City, Iowa]). Woolman, 1896: 352 (Little Minnesota R., Browns Valley, Minn. [and S. D.]).
Alvordius maculatus. Churchill and Over, 1933: 78, fig. 69 (description, importance; Whetstone Cr.).
Percina maculata. Underhill, 1957: 22, map 25 (Little Minnesota R., [Roberts Co.]).

The blackside darter is common in the Minnesota River drainage. Elsewhere it is known in South Dakota only from Meek's report in the Big Sioux River, a record that receives confirmation from Cleary's (1956: 319) and Underhill's (1957: 22) records from the Rock River in Lyon County, Iowa, and Rock County, Minnesota, respectively. Evermann and Cox (1896: 420) listed the species from Norfolk Creek at Norfolk Junction, and the Elkhorn River at Ewing, the only known localities in Nebraska (Johnson, MS). Woolman (1896: 360) recorded *Etheostoma aspro* from the James River, Jamestown, North Dakota, and the blackside darter is common in the Red River drainage of North Dakota and Minnesota.

Station records: 8, 9.

89. *Percina phoxocephala* (Nelson)—Slenderhead darter

The slenderhead darter, not previously known from South Dakota, was taken in Whetstone Creek of the Minnesota River drainage. Elsewhere in that drainage in Minnesota it is well distributed (Underhill, 1957: 22), but it apparently reached the area too late to utilize the postglacial Warren waterway into the Red River basin. The single South Dakota collection, UMMZ 166906, from station 7, consists of 7 specimens 30 to 61 mm. in standard length.

90. *Percina caprodes semifasciata* (DeKay)—Logperch

Percina caprodes. Churchill and Over, 1933: 76–77, fig. 66 (description, ecology, importance; larger lakes of eastern part of state).

Although the logperch occurs in the Minnesota and Red River basins in Minnesota (Underhill, 1957: 22, 27) our only South Dakota records are in the Big Sioux drainage, to which it gained access from the Minnesota or Des Moines drainages (see p. 123). Although our only specimens come from Lake Kampeska, Allum has seen specimens from Lake Cochrane, Deuel County; Lake Poinsett, Hamlin County; and either Lake Madison or Brant Lake, Lake County, and the logperch probably lives in a number of additional lakes of the Big Sioux drainage.

Station records: 15, 16.

KEY TO SPECIES OF *Etheostoma*

1a.—Anal spine single, thin, and flexible. Premaxillae protractile. [Interpelvic space wide, at least three-fourths of fin base.] .... Johnny darter, *Etheostoma nigrum*
1b.—Anal spines two, the first heavy and stiff. Premaxillae bound to snout by a frenum (rarely crossed by a groove in *zonale*). 2
 2a.—Pelvic fins widely spaced, the interspace about three-fourths of fin base. Lateral line complete. Pectoral fin longer than head. Snout very blunt. [Gill membranes broadly joined. Cheek scaled.] (*Hypothetical in South Dakota.*) ....
 .... Banded darter, *Etheostoma zonale*
 2b.—Pelvic fins more closely approximated, the interspace less than two-thirds of fin base. Lateral line incomplete. Pectoral fin shorter than head. Snout more or less sharp, not steeply declivous. 3
  3a.—Gill membranes separate or narrowly united. Dorsal spines usually 9 or more, their tips not thickened. Head with some scales. Infraorbital canal complete. .. 4
   4a.—Cheek well scaled. Supratemporal canal incomplete. Transverse scale rows 52 to 67. .... Iowa darter, *Etheostoma exile*
   4b.—Cheek naked or with a few scales behind eye. Supratemporal canal complete. Transverse scale rows 37 to 46. (*Hypothetical in South Dakota.*) ....
    .... Rainbow darter, *Etheostoma caeruleum*
  3b.—Gill membranes broadly united across isthmus to form a gentle curve. Dorsal spines 7 to 9, often (in adults) with thickened fleshy tips. Head scaleless. Infraorbital canal incomplete. (*Hypothetical in South Dakota.*) ....
   .... Fantail darter, *Etheostoma flabellare lineolatum*

91. *Etheostoma nigrum* (Rafinesque) —Johnny darter

Etheostoma nigrum. Meek, 1892: 246 (Big Sioux R., Sioux Falls, S. D., and Sioux City, Iowa [and S. D.]). Woolman, 1896: 352–53, 358 (description; Big Stone L., Creagers farm and Ortonville, Minn. [and S. D.]; Little Minnesota R.*, Browns Valley, Minn. [and S. D.]; Wheatstone [Whetstone] Cr., Milbank).
Boleosoma nigrum. Evermann and Cox, 1896: 421 (description; Enemy and Rock crs., Mitchell). Churchill and Over, 1933: 77, fig. 67 (rivers and creeks, especially east of Missouri R.).

Etheostoma nigrum nigrum. Cleary, 1956: 321 (Big Sioux R., Lyon [Lincoln] and Sioux [Lincoln and Union] cos., Iowa [and S .D.]). Underhill, 1959: 102 (Vermillion R. basin).

Etheostoma nigrum is known in South Dakota from the James, Vermillion, Big Sioux, and Minnesota river drainages; it should be sought also in the Red River drainage since it is widespread elsewhere in that basin. In Nebraska the species appears to be absent from Missouri River tributaries above the mouth of the Platte, although it occurs in the Platte system (Johnson, MS). It seems likely that *E. nigrum,* like several other species (p. 123). entered the middle Missouri basin from the Minnesota or the Des Moines basin.

All series of *Etheostoma nigrum* from South Dakota are structurally intermediate between subspecies *eulepis* and *nigrum* (see Lagler and Bailey, 1947); that is, the nape, cheek, and breast are usually neither naked nor fully invested with scales, but are partly scaled. Some individuals favor one subspecies or the other, but most are clearly intermediate (Table 8). Since the population in Big Stone Lake and its tributaries is like those in the Missouri basin, this morphological evidence supports the zoogeographic interpretation (pp. 124–25). It is probably not coincidental that *Etheostoma nigrum* exhibits essentially the same distributional pattern in South Dakota as do *Notropis stramineus stramineus, Notropis hudsonius, Rhinichthys atratulus,* and *Esox lucius.*

Station records: 3, 4, 5, 7, 8, 9, 15, 16, 18, 19, 22, 23, 24, 25B, 26C, 27, 28, 29, 31, 36, 39, 41, 43, 44.

92. *Etheostoma exile* (Girard) —Iowa darter

Etheostoma iowae. Woolman, 1896: 358 (Wheatstone [Whetstone] Cr., Milbank; [Little Minnesota R., Browns Valley, Minn., and S. D., UMMZ 177405; not recorded]). Evermann and Cox, 1896: 421–23 (description, habitat, comparison; Crow Cr.*, [near] Chamberlain; Enemy and Rock crs., Mitchell; Emanuel and Choteau* crs., Springfield; Prairie Cr., Scotland).

Oligocephalus iowae. Churchill and Over, 1933: 77–78, fig 68 (description; most of lakes and streams of eastern half of state).

Etheostoma exile. Underhill, 1957: map 28 (Lake Traverse and Little Minnesota R. [Roberts Co.]; Big Stone L. [Grant Co.]). Moyle and Clothier, 1959: 179 (Lake Traverse). Underhill, 1959: 102 (Vermillion R., gravel riffles south of Centerville).

Etheostoma exile lives in all principal drainages east of the Missouri River in South Dakota and in the Niobrara drainage to the west. The species is partial to clear, sluggish, or standing waters with vegetation, and restriction of this habitat is doubtless the chief factor limiting its occurrence in the northern plains area. The Iowa darter ranges through North Dakota, eastern Montana, and parts of the Prairie provinces of Canada. It is appar-

TABLE 8

FREQUENCY DISTRIBUTIONS OF ESTIMATED PERCENTAGES OF SQUAMATION OF NAPE, CHEEK, AND BREAST IN *Etheostoma nigrum* FROM SOUTH DAKOTA

Drainage and Locality	Sta. No.	0	10	20	30	40	50	60	70	80	90	100	N	M
						NAPE								
Minnesota R.:														
Big Stone L.	3, 4	..	..	..	2	1	..	..	..	..	2	4	9	75.6
Whetstone Cr.	7, 8	3	4	8	7	3	2	4	5	8	6	..	50	48.0
Yellowbank R.	9	1	4	6	4	..	1	2	..	1	4	2	25	43.6
Big Sioux R.:														
Six-mile Cr.	25	4	2	3	3	1	..	..	2	2	..	..	17	30.0
James R.:														
James R.	43	..	..	1	2	2	..	1	..	..	..	..	6	36.7
						CHEEK								
Big Stone L.	3, 4	1	..	..	1	..	1	1	1	1	3	..	9	62.2
Whetstone C.	7, 8	9	8	1	8	2	4	5	7	5	1	..	50	38.0
Yellowbank R.	9	5	3	5	3	..	..	3	4	2	..	..	25	30.8
Six-mile Cr.	25	3	3	2	3	..	2	..	4	..	..	..	17	31.8
James R.	43	3	..	2	1	..	..	..	..	..	..	..	6	11.7
						BREAST								
Big Stone L.	3, 4	..	..	1	..	1	..	..	1	1	4	1	9	74.4
Whetstone Cr.	7, 8	1	..	4	4	9	8	7	4	9	2	2	50	55.2
Yellowbank R.	9	..	4	3	3	5	2	..	5	2	1	..	25	43.6
Six-mile Cr.	25	..	..	..	4	..	2	1	..	8	2	..	17	64.7
James R.	43	..	..	1	1	4	..	..	..	..	..	..	6	35.0
					MEAN OF NAPE, CHEEK, AND BREAST [*]									
Big Stone L.	3, 4	..	..	1	1	..	..	1	1	..	4	1	9	71.1
Whetstone Cr.	7, 8	1	4	9	6	4	5	4	7	9	1	..	50	47.0
Yellowbank R.	9	..	7	3	4	2	1	..	3	4	1	..	25	40.0
Six-mile Cr.	25	..	2	3	2	3	3	..	4	..	..	..	17	40.6
James R.	43	..	..	3	2	1	..	..	..	..	..	..	6	26.7

[*] Rounded to nearest 10 per cent.

ently much less common in South Dakota now than when the state was studied by Evermann and Cox (1896: 421–22).

Station records: 1, 3, 6, 11, 14, 19, 23, 25B, 27, 31, 36, 37, 38, 39, 41, 43, 44, 53, 55, 57, 84, 85.

SCIAENIDAE—DRUMS

93. *Aplodinotus grunniens* Rafinesque—Freshwater drum

Aplodinotus grunniens. Meek, 1892: 246 (Big Sioux R.*, Sioux City, Iowa [and S. D.]). Cox, 1896: 611 (Little Minnesota R., Browns Valley, Minn. [and S. D.] and Big Stone L.). Woolman, 1896: 353 (Little Minnesota R., Browns Valley, Minn. [and S. D.]; Big Stone L., Creagers farm and Ortonville, Minn. [and S. D.]). Churchill and Over, 1933: 79, fig. 70 (description, ecology, importance; larger rivers of eastern half of state and in some lakes). Cleary, 1956: 324 (Big Sioux R., Lyon [Lincoln] and Plymouth [Union] cos., Iowa [and S. D.]). Shields, 1958a: 31; 1958b: 360 (Ft. Randall Res.). Hugghins, 1959: 23 (Big Stone L.). Underhill, 1959: 102 (lower reaches of Vermillion R. and at mouth).

Aplodinotus grunniens is an inhabitant of large waters. It entered South Dakota through two channels, in Big Stone Lake, from which it has spread into the Red River basin, and in the Missouri River, in which it ranges upstream into North Dakota and Montana. Except for these waters, where it is common, the freshwater drum is known to occur only in the lower parts of the Big Sioux, Vermillion, and James rivers.

Station records: 3, 4, 8, 51, 64, 65, 72, 76.

ADDITIONAL SPECIES REPORTED OR OF HYPOTHETICAL OCCURRENCE IN SOUTH DAKOTA, INCLUDING UNACCEPTABLE RECORDS

Ichthyomyzon castaneus Girard.—The chestnut lamprey was mapped as occurring in South Dakota by Hubbs and Trautman (1937: 44) on the basis of the record of *Ichthyomyzon concolor* (Evermann and Cox, 1896: 384) from Crow Creek near Chamberlain, Buffalo County. Hubbs and Trautman did not examine the specimen, and doubt has recently been raised as to the identification of South Dakota lampreys by the discovery that *Ichthyomyzon unicuspis* formerly occurred in the Missouri River at Vermillion (Bailey, 1959b: 163). Until clear evidence for the occurrence of any other lamprey in the Missouri drainage of the state is obtained, it seems best to assume that unverified specimens also are *unicuspis*. *I. castaneus* should be looked for in Big Stone Lake.

Acipenser fulvescens Rafinesque.—The lake sturgeon (as *Acipenser rubicundus* LeSueur) was reported on hearsay evidence to occur in the White and Missouri rivers near Chamberlain and near Yankton by Evermann and Cox (1896: 385). No basis for the identification was given, but large size presumably contributed to the determination. The discovery that *Scaphirhynchus albus* occurs in the upper Missouri and that it attains a large size (Bailey and Cross, 1954: 201–202; Brown, 1955: 55) suggests that this species provided the basis for the report. We know of no verified record of the lake sturgeon in the Missouri basin above the mouth of the Kansas River (Johnson, MS). We have heard unconfirmed reports of the former occurrence of the lake sturgeon in Big Stone Lake.

Salmo clarki Richardson.—Suckley (1874: 139) reported cutthroat trout, as *S. lewisi*, from the Black Hills of "Nebraska." Evermann and Cox (1896: 384) were unable to verify this reference and believed it to be erroneous. Furthermore, Captain William Ludlow, of Custer's expedition to the Black Hills in 1874, reported (1875: 16): "We were continually looking for trout in these streams . . . we found nothing but some small chub [sic], and a species of sucker. . . ." We know of no support for the past or present occurrence of cutthroat trout in South Dakota.

Hiodon tergisus LeSueur.—*Hyodon tergisus* was reported from the upper Missouri basin by Evermann and Cox (1896: 413). The only South Dakota record (Cope, 1879: 441) is from the lower part of Battle Creek [perhaps now known as Blue Blanket Creek, below Mobridge, Walworth County]. However, Cope did not list *Hiodon alosoides* from this or other localities and we therefore assume that his records of the mooneye were based on misidentifications of the abundant goldeye. We know of no records that justify the inclusion of the mooneye in faunal lists from the upper Missouri basin. The species should be sought in the Minnesota River drainage.

Hybopsis aestivalis (Girard).—The speckled chub has not been taken in the Mississippi River drainage above Vernon County, Wisconsin (Greene, 1935: 79), or in the Missouri River drainage above the mouth of the Platte River. The Museum of Zoology has Nebraska collections taken by Raymond E. Johnson between 1939 and 1941 in the Platte River (Hall and Butler cos.), Elkhorn River (Dodge Co.), Loup River (Platte Co.), South Loup River (Buffalo Co.), and North Loup River (Greeley Co.), as well as lots from several stations on the Republican River. The Iowa State University collection has a specimen (ISU 1664) taken in the Missouri River, one mile below the mouth of the Platte River, Mills County, Iowa, by J. R. and R. M. Bailey. It seems unlikely that the species occurs in South Dakota.

Notropis spilopterus (Cope).—Meek (1892: 246) reported *Notropis whipplei*, which was formerly confused with the spotfin shiner, *N. spilopterus*, from the Big Sioux River at Sioux City, Iowa. *N. spilopterus* is an abundant species of the upper Mississippi drainage, but is apparently absent from the Missouri basin (Gibbs, 1957: 194) except in southern Missouri and in the Spirit Lake area of northwestern Iowa (where presumably introduced; Meek, 1894, and Larrabee, 1926, did not take it in early surveys). Records from western Iowa such as those of Meek (*op. cit.*) and Cleary (1956: 299) are based on misidentifications of *Notropis lutrensis* or on recent introductions. Meek failed to report *lutrensis* from the Big Sioux River although it is now common there.

It is possible that *Notropis spilopterus* occurs in the Minnesota River drainage of South Dakota since it has been reported by Underhill (1957: map 19) from the Minnesota River and tributaries only a few miles below Big Stone Lake. Until a firm record of occurrence is available the species may be omitted from the South Dakota faunal list.

Notropis anogenus Forbes.—*Hybopsis anogenus* was reported by Churchill and Over (1933: 49–50, fig. 36) from "the lakes and streams east of the Missouri River." No specific localities were given and no specimens from the state are known to be present in collections. The figure appears to be redrawn from that of Forbes and Richardson (1909: 132, fig. 29). The pugnose shiner has been collected in eastern North Dakota and in northwestern Iowa (Bailey, 1959*a*) so it possibly occurs or once did live in South Dakota (see p. 123), but until an actual verified record is available it seems best to delete it from the state faunal list. The species inhabits clear lakes and sluggish waters with much vegetation.

Pimephales vigilax perspicuus (Girard).—The bullhead minnow, as *Ceratichthys perspicuus*, was mapped by Hubbs and Black (1947: 30) as occurring in eastern Nebraska, western Iowa, and in both the Missouri and Minnesota river drainages of South Dakota. With the exception of an apparently erroneous locality in western Iowa (Hubbs, personal communication), the record stations were taken entirely from the literature (Evermann and Cox, 1896: 400; Churchill, 1927; Churchill and Over, 1933: 42–43), and for some the accuracy was questioned (Hubbs and Black, 1947: 28). Larrabee (1926) suspected that Iowa specimens from the Missouri drainage had escaped from bait buckets. Dr. Raymond E. Johnson reexamined the Nebraska specimens of Evermann and Cox and found them to be *Pimephales notatus*. Recent surveys of Nebraska (Johnson), Iowa (Cleary, 1956), Minnesota (Underhill, 1957), and South Dakota (this paper) have yielded no new records of the bullhead minnow from the Missouri or Minnesota river drainages of these states. We agree with Hubbs (1951:8) that the old records are unacceptable; their elimination results in a considerable modification of the range of the species (Hubbs and Black, 1947: 30).

Moxostoma valenciennesi Jordan.—A specimen of the greater redhorse (USNM 68215) has been recorded from the Maple River, North Dakota, by

Robins and Raney (1957: 154). This fish was collected by Woolman and is the only one known from the Red River drainage. The species may occur in the Red or Minnesota river system in northeastern South Dakota.

Moxostoma carinatum (Cope).—A specimen of the river redhorse from the S. E. Meek collection (now UMMZ 162877, 150 mm. in standard length) bears the data "Floyd River near Sioux City, Iowa"; this confirms Meek's (1894: 136) identification. This locality is remote from the general range of the species, but if the data are correct a wider former distribution is indicated. Recent restriction of range is apparently correlated with in-increased pollution and turbidity (Trautman, 1957: 262). The river redhorse may once have lived in southeastern South Dakota, but it presumably does not at present.

Moxostoma anisurum (Rafinesque).—Although we know of no South Dakota record for the silver redhorse, Eddy and Surber (1947: 135–36) reported it as abundant in the Minnesota River, and its discovery in Big Stone Lake may be anticipated.

Moxostoma duquesnei (LeSueur).—The reallocation of names of the redhorses (Hubbs, 1930) renders hazardous the acceptance of earlier records unless these are confirmed by reidentification of specimens. Hence, reports of this species from the Big Sioux River at Sioux Falls, South Dakota, and Sioux City, Iowa (Meek, 1892: 246) are inadmissible. They probably refer to the northern redhorse, *Moxostoma macrolepidotum,* the common redhorse of the area.

Ictalurus natalis (LeSueur).—There are no South Dakota records for the yellow bullhead, but it may be regarded as a likely future addition to the state faunal list. It has been reported from the James River at La Moure, North Dakota (Woolman, 1896: 359), from near Sioux City, Woodbury County, Iowa (Cleary, 1956: 306), and from various localities in southern Minnesota (Eddy and Surber, 1947: 178).

Aphredoderus sayanus (Gilliams).—The pirate perch was stated to range into South Dakota by Hubbs and Lagler (1941: 68) in their guide to Great Lakes fishes. This report is erroneous and has been corrected in subsequent editions of their work.

Lepomis megalotis (Rafinesque).—As indicated by Hubbs (1945: 20–21) the records of *Lepomis megalotis* from Big Stone Lake at Ortonville, Minnesota, and Wheatstone [sic] River at Millbank [sic], South Dakota (Woolman, 1896: 352, 358) are suspect. They almost certainly are based on *Lepomis humilis,* a not uncommon species of the region that was not listed by Woolman. The only records of the longear sunfish from Minnesota (or North or South Dakota) known to us are those reported by Eddy and Surber (1947: 237–38) from Little Rock Lake, Morrison County, and an individual collected by Raymond E. Johnson on August 23, 1945, in Trout Lake, T. 57–58 N, R. 25 W, Itasca County, both in the upper Mississippi River drainage. The latter was identified by Bailey on February 27, 1947, as *Lepomis megalotis peltastes* Cope. It has 35 or 36 lateral-line scales, 17 rows of caudal-peduncle scales and 12–13 pectoral rays, characters that contrast with *L. m. megalotis*. The specimen is in the collection of the Minnesota Department of Conservation.

Percina shumardi (Girard).—Since the river darter occurs in the Mississippi River (Greene, 1935: 172) and in the Red River of the North (Hankinson, 1929: 454), its use of the Warren glacial outlet is indicated. Despite an absence of South Dakota records the species may occur in Big Stone Lake.

Etheostoma zonale (Cope).—The banded darter was reported as rare in the Big Sioux River, Sioux City, Iowa, by Meek (1892: 246), a record that may rest on misidentification, perhaps of *Etheostoma exile*. There is no other record of *E. zonale* from the Missouri River drainage except in the Ozark upland of southern Missouri. But *Etheostoma zonale* occurs in the Minnesota River drainage of Minnesota within about 20 miles of South Dakota (Underhill, 1957: 23, map 27), and should be searched for in adjacent South Dakota.

Etheostoma caeruleum Storer.—The rainbow darter has not been found in South Dakota, but since it occurs (together with *E. zonale* and *E. flabellare lineolatum*) in the Minnesota River drainage close by (Underhill, 1957: 24, map 29) it may live in the state.

Etheostoma flabellare lineolatum (Agassiz).—Like the two preceding species, the fantail darter lives in western Minnesota (Underhill, 1957: 24, map 30) and should be looked for in South Dakota.

HYBRIDIZATION

That interspecific hybrids are of frequent natural occurrence among North American freshwater fishes has been amply demonstrated by Dr. Carl L. Hubbs and others during recent decades. Most often such hybrids probably make no reproductive contribution and thus play no positive role in evolution. Their study, however, may permit assessment of the genetic basis for species differences, and thus contribute to character evaluation in phyletic inquiry. From a practical viewpoint hybrids pose problems of identification, and failure to identify them accurately may lead to the accumulation of inaccurate zoogeographic or ecologic data. Several nominal species created years ago are now known to rest on hybrid individuals. The hybrid combinations encountered in this study are listed herewith.

Campostoma anomalum × *Rhinichthys cataractae*

A specimen (UMMZ 127468, 49 mm. standard length) taken at station 84 was associated with both parental species.

Chrosomus eos × *Chrosomus neogaeus*

Four specimens (UMMZ 127452, 38 to 55 mm.) were collected at station 85. Three individuals of *C. eos* were obtained in the same collection, but none of *C. neogaeus* was caught either here or elsewhere in the area.

Hybopsis biguttata × *Notropis cornutus*

A single individual (UMMZ 166895, 48 mm.) taken at station 7 was associated with both parent species.

Catostomus commersoni × *Pantosteus platyrhynchus*

Catostomus commersonnii sucklii × *Pantosteus jordani.*—Hubbs, Hubbs, and Johnson, 1943: 37–39 (characters; records from stations 106 and 108 of this paper).

Lepomis cyanellus × *Lepomis gibbosus*

The collection at station 37 included three half-grown hybrids, three green sunfish and two pumpkinseeds.

Lepomis cyanellus × *Lepomis humilis*

Single specimens of this combination were taken together with both parental species at stations 34 (UMMZ 163228, 71 mm.) and 133.

Pomoxis annularis × *Pomoxis nigromaculatus*

A young hybrid crappie (UMMZ 127423, 34 mm.) was taken at station 51 together with 27 specimens of *annularis* and 10 of *nigromaculatus*.

SOME HYDROGRAPHIC INTERCHANGES THAT AFFECT FISH DISTRIBUTION IN SOUTH DAKOTA

The dispersal of South Dakota fishes mainly involves established modern watercourses, but three exceptions require comment. The geological history of the postglacial water connections between the drainages of the Arctic Ocean and the Gulf of Mexico is well documented, and its ichthyological consequence has been noted. Intimate hydrographic connections exist also between the upper Mississippi River basin and the middle Missouri drainage, and their recognition, although apparently not appreciated previously by ichthyogeographers, is essential to an understanding of fish distribution in the area. The beheading of the Little Missouri River in northeastern Wyoming by the Belle Fourche River is a readily demonstrable geological fact, but whether or not it has important biological consequences is uncertain.

RIVER WARREN.—During glacial withdrawal, Lake Agassiz, largest of glacial lakes (Upham, 1896) developed in northcentral United States and southern Canada. It was blocked to the north by a glacial dam and discharged to the south through River Warren, which flowed through the valley now occupied by Lake Traverse, Big Stone Lake, and the Minnesota River, to the Mississippi River. Fish utilized this outlet as a pathway for northward redispersal into the Arctic drainage (Greene, 1935: 12–15). With the eventual withdrawal of the ice a lower outlet was freed in the north and Lake Agassiz was largely drained. At Browns Valley, a village in the floor of the ancient valley between Lake Traverse and Big Stone Lake, there developed a low divide between the drainages of the Arctic Ocean and the Gulf of Mexico (Leverett, 1932: 123–26; Flint, 1955: 127). The divide lies only about three feet above the average level of Lake Traverse and eleven feet above that of Big Stone Lake (Upham, 1896: 17–18), but it is an effective modern barrier to fish dispersal (Underhill, 1957: 6–7) that seems not to be rendered ineffective by occasional spring flooding.

UPPER MISSISSIPPI AND MIDDLE MISSOURI.—The fish faunas of the upper Mississippi and the Missouri rivers share many elements, but as this study proceeded we were confronted with the appearance of many forms in the middle Missouri basin which, on the basis of general distribution, seem out of place. With few exceptions these are absent from the lower part of the Missouri drainage. We suspected that the glacial history of the area would reveal a major, temporary postglacial channel, perhaps involving a northern waterway between the James and Red river drainages. A canvass of the geological evidence (e. g., Upham, 1896; Leverett, 1932; Flint, 1955; Lemke, 1960), however, rather effectively contradicted this hypothesis. Similarly, the steep eastern slope of the Coteau des Prairies, separating the upper Minnesota and Big Sioux drainages in northeastern South Dakota (Flint, 1955: pl. 2), appears not to offer a suitable route. Study of topographic maps to the southeast, by contrast, reveals several points at which stream topography is strongly suggestive of hydrographic intercommunication, stream capture, or flood connections between tributaries to the Big Sioux and Little Sioux rivers of the Missouri basin on one side and the Minnesota and Des Moines rivers of the upper Mississippi basin on the other.

Lake Hendricks, at an elevation of 1730 feet, astride the line between Brookings County, South Dakota, and Lincoln County, Minnesota, is shown on the Watertown sheet (1958, U. S. Geol. Surv., map NL14–12; scale 1:250,000) draining to the east into the Lac Qui Parle River of the Minnesota River, and having a western connection with Deer Creek of the Big Sioux drainage. On inquiry, we learn from State Game Warden Elmer Liebig, Brookings, South Dakota, that the chief tributary to Lake Hendricks divides during heavy floods, pouring part of its waters northward for a distance of a mile into the southwest corner of the lake, whereas the bulk of the flood flow continues southward into Deer Creek. Although Lake Hendricks does not flow into Deer Creek there is only a four-foot rise in the mile separating the lake and the point of overflow on the creek. Fish could easily travel from the lake to Deer Creek or vice versa. Mr. Liebig has observed this flood phenomenon several times during the past twenty years, the last time in the spring of 1960.

Additional areas of possible intercommunication include Lake Benton, elevation 1730 feet, Lincoln County, Minnesota (Watertown sheet), shown draining to the northeast to the Redwood River of the Minnesota River. A narrow divide on a possible former watercourse one mile southwest of the lake is continuous with Flandreau Creek, a tributary to Big Sioux River. Leverett (1932: pl. 2) shows Flandreau Creek as the outlet from Lake Benton. Okabena Lake, elevation 1568 feet, near Worthington, Nobles County, Minnesota (Fairmont sheet, 1958, U. S. Geol. Surv., map NK15–1), drains through Okabena Creek to the West Fork of the Des Moines River. Another creek one-half mile distant to the southeast (one mile south of Worthington) and without intervening contour (interval, 50 feet) flows south for 1.5 miles into Ocheda Lake, elevation 1565 feet, tributary to the Ocheyedan River of the Little Sioux River. Leverett (1932: pl. 2) shows this as the outlet route from Okabena Lake. Still another creek only one-half mile from the southwest corner of Okabena Lake flows south to Ocheda Lake.

It seems clear that the low divides and many waterways and lakes of the moraines of southwestern Minnesota have provided Recent channels for the movement of fishes between the upper Mississippi and Missouri basins. Study of the fishes believed to have utilized these connectives reveals them to be ecologically well suited to the waters of the area, and most are known to occur here.

BELLE FOURCHE AND LITTLE MISSOURI RIVERS.—Geologically, the piracy of the former headwater part of the Little Missouri River by the Belle Fourche River is classic in its simplicity. At present the Belle Fourche rises in the southwest part of Campbell County, Wyoming, and follows a northeast course for over 110 miles (straight line) to a point northwest of

Colony in northeastern Crook County, Wyoming (T. 57 N, R. 62 W, sec. 7), where it turns abruptly to the southeast, a course followed with minor deviation to its confluence with the Cheyenne River. The site of the abrupt change in course, and of stream capture, is on the Alladin Topographic Quadrangle (U. S. Geol. Surv., scale 1:125,000), at an altitude of about 3360 feet. From just above the dissected stream bed here, a flat valley floor, between 3450 and 3500 feet high, extends north-northwest for a distance of about six miles to the Little Missouri River, which in this section has a rather low gradient and is about 3450 feet high. When the rapidly eroding headwaters of the Belle Fourche River cut back into the Little Missouri that stream was beheaded and its upper part with its aquatic inhabitants was added to the Belle Fourche River. Although the timing is not known to us, the geological reality of the stream transfer is evident. What faunal shifts were involved is uncertain. The juxtaposition of the range of *Catostomus catostomus* in South Dakota with this area is highly suggestive. At present the upper Belle Fourche and the Little Missouri apparently do not provide a suitable habitat for that species, but this does not preclude past existence there. *Pantosteus platyrhynchus* and *Hybopsis plumbea* live also in the Black Hills area; both thrive in the upper Missouri basin and the latter survives in the Little Missouri. For these species, however, there are other records of occurrence in the middle Missouri basin (see pp. 43, 87). There is some evidence that the Black Hills stock of *Rhinichthys cataractae* may be derived from the upper Missouri basin (see p. 53).

ORIGIN AND COMPOSITION OF THE SOUTH DAKOTA FISH FAUNA

Hydrographically South Dakota comprises three major drainages (Fig. 1). (a) The Missouri basin includes 97 per cent of the area of the state and has a known native fish fauna of 76 forms (Table 9). These include a few glacial relicts, many forms that entered the basin postglacially from the Missouri River, and a component derived by hydrographic interchange with the Upper Mississippi River basin. (b) The Minnesota River drainage occupies about two per cent of the state and contains 50 native fishes. All probably gained access to the basin by northward and westward dispersal in the upper Mississippi drainage. (c) The Red River drainage occupies less than one per cent of the area of the state and is known to have 18 native fishes, only a fraction of the total number in the basin. All apparently utilized the River Warren in northward postglacial dispersal from the upper Mississippi drainage (p. 113).

RELICT SPECIES.—Of the 94 species and subspecies of fishes here listed

TABLE 9

Sources of Origin and Distribution by Hydrographic Areas of South Dakota Fishes

Source of Origin: 1 = preglacial relicts in or adjacent to western South Dakota; 2 = Missouri River drainage; 3 = Upper Mississippi River drainage; 4 = artificial introduction. Occurrence: X = specimens examined by authors or Carl L. Hubbs; R = acceptable published reports; I = introductions of unknown success; ( ) = known or suspected introductions.

Family and Species	Source of Origin	Red R. dr.	Minnesota R. dr.	Big Sioux R. dr.	Vermillion R. dr.	James R. dr.	E tribs. Missouri R.	Missouri R.	SW tribs. Missouri R.	Cheyenne R. dr.	NW tribs. Missouri R.	Little Missouri R. dr.
Petromyzontidae												
1. *Ichthyomyzon unicuspis*	2						R?	X				
Acipenseridae												
2. *Scaphirhynchus platorynchus*	2							X				
3. *Scaphirhynchus albus*	2							X				
Polyodontidae												
4. *Polyodon spathula*	2			R				X				
Lepisosteidae												
5. *Lepisosteus platostomus*	2					R	X	R?	X			
6. *Lepisosteus osseus*	2,3		R	R				R?	X			
Amiidae												
7. *Amia calva*	3		R?	?								
Clupeidae												
8. *Alosa chrysochloris*	2,3		R					X				
9. *Dorosoma cepedianum*	2			X	R	X	X	X				
Salmonidae												
10. *Salmo trutta*	4								(X)	(X)	(I)	
11. *Salmo gairdneri*	4	(X)	(X)			(I)		(I)	(I)	(X)	(I)	
12. *Salvelinus fontinalis*	4	(I)								(X)		
Umbridae												
13. *Umbra limi*	3		X									
Esocidae												
14. *Esox lucius*	3	R	X	X	(R)	X	(I)			(I)		
Hiodontidae												
15. *Hiodon alosoides*	2			R	R	X	X	X		R	X	X

TABLE 9 (Continued)

Family and Species	Source of Origin	Red R. dr.	Minnesota R. dr.	Big Sioux R. dr.	Vermillion R. dr.	James R. dr.	E tribs. Missouri R.	Missouri R.	SW tribs. Missouri R.	Cheyenne R. dr.	NW tribs. Missouri R.	Little Missouri R. dr.
Cyprinidae												
16. *Cyprinus carpio*	4	(X)	(X)	(X)	(X)	(X)	(X)	(X)	(X)	(X)	(X)	(X)
17. *Carassius auratus*	4							(X)	(R)	(X)		
18. *Notemigonus crysoleucas*	2,3		R	X	X	X			(X)	(X)		
19. *Semotilus margarita*	1								X			
20. *Semotilus atromaculatus*	2,3	R	X	X	X	X	X		X	X	X	
21. *Chrosomus neogaeus*	1									X		
22. *Chrosomus eos*	1,3		X				R		X			
23. *Hybopsis plumbea*	1						R			X		X
24. *Hybopsis gracilis*	2			R	R				X	X	X	X	X
25. *Hybopsis biguttata*	3	R	X	R								
26. *Hybopsis storeriana*	2							X				
27. *Hybopsis gelida*	2								X	X	X	X
28. *Hybopsis meeki*	2								X			
29. *Rhinichthys atratulus*	3		X	X	R	X	X					
30. *Rhinichthys cataractae*	1,2,3?	R?							X	X	X	X
31. *Phenacobius mirabilis*	2						R					
32. *Notropis atherinoides*	2,3		X	X	R	X		X				
33. *Notropis rubellus*	3		X									
34. *Notropis illecebrosus*	2							X				
35. *Notropis cornutus*	2,3	R	X	X	R	X	R					
36. *Notropis heterodon*	3		R									
37. *Notropis hudsonius*	3		X	X		X						
38. *Notropis blennius*	2							R				
39. *Notropis dorsalis*	2,3	R	X	X	R	X	X	X	X			
40. *Notropis lutrensis*	2			X	X	X	X	X				
41. *Notropis stramineus*												
41a. *N. s. stramineus*	3		X	X	X	X						
41b. *N. s. missuriensis*	2							X	X	X	X	X
42. *Notropis topeka*	2			X	X	X						
43. *Notropis heterolepis*	2,3	R	R	R			X	R				
44. *Hybognathus hankinsoni*	2,3	X	X	X	X	X	X		X			X
45. *Hybognathus placitus*	2							X	X	X	X	X
46. *Hybognathus nuchalis*	2			R	R			X	X	X	R	X	X
47. *Pimephales notatus*	3	R	X	X						(X)		
48. *Pimephales promelas*	2,3	X	X	X	X	X	X	X	X	X	X	X
49. *Campostoma anomalum*	2,3		X	X	R	X	R		X			

TABLE 9 (Continued)

Family and Species	Source of Origin	Red R. dr.	Minnesota R. dr.	Big Sioux R. dr.	Vermillion R. dr.	James R. dr.	E tribs. Missouri R.	Missouri R.	SW tribs. Missouri R.	Cheyenne R. dr.	NW tribs. Missouri R.	Little Missouri R. dr.
Catostomidae												
50. *Cycleptus elongatus*	2			R				X				
51. *Ictiobus cyprinellus*	2,3		X	X	X	X		X				
52. *Ictiobus bubalus*	2			R	R	X		X				
53. *Carpiodes cyprinus*	2,3		X	R								
54. *Carpiodes carpio*	2			X	X	X	X	X		X	X	X
55. *Hypentelium nigricans*	3		R									
56. *Moxostoma erythrurum*	3		X									
57. *Moxostoma macrolepidotum*	2,3		X	R	R	X	X		X	X	X	X
58. *Catostomus commersoni*	2,3	R	X	X	R	X	X	X	X	X	X	X
59. *Catostomus catostomus*	1									X		
60. *Pantosteus platyrhynchus*	1									X		
Ictaluridae												
61. *Ictalurus melas*	2,3	X	X	X	X	X	X	X	X	X	X	X
62. *Ictalurus nebulosus*	3	R	R									
63. *Ictalurus punctatus*	2,3	R		R	R	X	R	X	X	X	X	X
64. *Ictalurus furcatus*	2								X			
65. *Noturus gyrinus*	2,3		X	X		X	R					
66. *Noturus flavus*	2,3		X	R	R			X	X	X	X	X
67. *Pylodictis olivaris*	2			R				R				
Anguillidae												
68. *Anguilla rostrata*	2,3			R	R	R						
Cyprinodontidae												
69. *Fundulus diaphanus*	3			X	R		(X)					
70. *Fundulus kansae*	4									(X)		
71. *Fundulus sciadicus*	2				X	X	X		X			
Gadidae												
72. *Lota lota*	2						X	X			R	
Gasterosteidae												
73. *Culaea inconstans*	2,3	R	X	X	X		R					
Percopsidae												
74. *Percopsis omiscomaycus*	3		R	X								
Serranidae												
75. *Roccus chrysops*	3		X	X								

TABLE 9 (Continued)

Family and Species	Source of Origin	Red R. dr.	Minnesota R. dr.	Big Sioux R. dr.	Vermillion R. dr.	James R. dr.	E tribs. Missouri R.	Missouri R.	SW tribs. Missouri R.	Cheyenne R. dr.	NW tribs. Missouri R.	Little Missouri R. dr.
Centrarchidae												
76. *Micropterus dolomieui*	3		R									
77. *Micropterus salmoides*	3	(X)	X	X	(R)	(X)	(X)	(X)	(X)	(X)	(X)	
78. *Lepomis cyanellus*	2			X	X	X	X	X	X	(X)	(X)	
79. *Lepomis gibbosus*	3		X	X		(X)			(X)		(X)	
80. *Lepomis macrochirus*	3	(X)	X	X	(R)	(X)	(X)	(R)	(X)	(R)	(X)	
81. *Lepomis humilis*	2,3	(X)	X	X	X	X	(X)	X	(X)	(X)	(X)	
82. *Ambloplites rupestris*	3		X	X							(X)	
83. *Pomoxis annularis*	2,3	(X)	X	X	(R)	(X)	(X)	(X)				
84. *Pomoxis nigromaculatus*	3	(X)	X	X	(R)	(X)	(X)	(X)	(X)		(X)	
Percidae												
85. *Stizostedion canadense*	2			X	R		R	X	R		X	
86. *Stizostedion vitreum*	2,3	R	X	X	(R)	R	R	X		(X)	X	
87. *Perca flavescens*	3	X	X	X	X	X	(X)	(X)	(X)	(X)	(X)	
88. *Percina maculata*	3		X	R								
89. *Percina phoxocephala*	3		X									
90. *Percina caprodes*	3				X							
91. *Etheostoma nigrum*	3		X	X	R	X						
92. *Etheostoma exile*	2,3	X	X	X	R	X	X		X			
Sciaenidae												
93. *Aplodinotus grunniens*	2,3		X	X	R	X		X				
Totals												
Native		18	50	58	35	34	33	38	23	20	17	15
Introduced		6	3	2	7	7	10	7	12	14	12	1
Combined		24	53	60	42	41	43	45	35	34	29	16

from South Dakota, at least six are believed to have persisted during the last glaciation in the western part of the state or in adjacent areas in the upper Missouri basin. Five of these are northern fishes, the present ranges of which extend into the Canadian Northwest Territories. Given moderate summer temperatures there is no reason to doubt their tolerance of Pleistocene conditions on the northern Plains. Each of these species has a disruptive distribution, with most of its range far to the north (Figs. 2 and 3). *Pantosteus platyrhynchus* is perhaps equally cold hardy, but its preference for swift water likely explains its failure to range far north on the central lowlands of Canada. Three cyprinids (*Semotilus margarita nachtriebi, Chrosomus neogaeus,* and *C. eos*) are common associates in isolated springs of the Sand Hills area of Nebraska and adjacent South Dakota, and *Hybopsis plumbea* is one of the dominant fishes of the Black Hills.

Two suckers, *Catostomus catostomus* and *Pantosteus platyrhynchus,* are confined in their distribution in South Dakota, to the Cheyenne River system, either in or near the Black Hills. Both occur also in the Upper Platte River and in the Upper Missouri basin. Perhaps these species gained access to the Cheyenne basin when the Belle Fourche River captured the former headwaters of the Little Missouri River (see p. 114). At present neither of these catostomids persists in the Little Missouri drainage (Personius and Eddy, 1955), but they are found in the Big Horn, Tongue, and Powder rivers (Evermann and Cox, 1896; Simon, 1946). Prior to the time of stream capture, both may have lived in the Little Missouri system. Alternatively, during cooler postglacial time they may have ranged far downstream in the Missouri whence they could have ascended the Cheyenne River to the Black Hills. Evermann and Cox (1896: 390) reported *Pantosteus* from the White River system in Chadron Creek, Chadron, Nebraska.

The six species of relict occurrence in South Dakota are: *Semotilus margarita nachtriebi, Chrosomus neogaeus, Chrosomus eos* (in western part of state), *Hybopsis plumbea, Catostomus catostomus,* and *Pantosteus platyrhynchus.*

In the Upper Missouri basin of Wyoming and Montana there are additional cool-water fishes that are similarly isolated, i.e., *Thymallus arcticus, Prosopium williamsoni, Salmo clarki,* and *Cottus bairdi.* It is likely that others, including *Rhinichthys cataractae,* remained in the Upper Missouri basin during the last glaciation, but for forms of wide distribution it becomes difficult to identify postglacial pathways of dispersal.

Preglacial drainage made the present Upper Missouri basin readily accessible to northern fishes since this area is believed to have drained northeastward into Hudson Bay (Fenneman, 1938: 564–66, fig. 152). The present southward flow of the Missouri dates only from the Wisconsin glacial stage.

MISSOURI RIVER.—As the largest river in the state and with drainage into the heart of the faunally rich Mississippi Valley, it is predictable that following glaciation the Missouri River should serve as a major pathway for entry of fishes into South Dakota. Of some 56 species that utilized this route, 28 employed it exclusively (Table 9). Thirty-three fishes may be classified as characteristic inhabitants of the Missouri and other large rivers, though some are not restricted to such waters. For these forms the river is an avenue of dispersal that has granted ready access to at least the lower courses of every principal stream system in the Missouri drainage of South Dakota. The present distribution of these species here is chiefly a function of their ecologic adaptability. Among these large-river species all, with the probable exception of *Lota lota,* occur downstream at or near the confluence of the Missouri and Mississippi rivers. (*Lota* prefers relatively cool water and doubtless occurred much farther downstream during the Pleistocene.) Among the Missouri River fishes, *Scaphirhynchus albus, Hybopsis gracilis, H. gelida, H. meeki,* and *Hybognathus placitus* are highly adapted to heavily silt-laden waters and none of these ascends the upper Mississippi appreciably above the mouth of the turbid Missouri River (Bailey and Cross, 1954). *Notropis illecebrosus* has a similar distribution but lives a short distance upstream in the Mississippi River (Gilbert and Bailey, MS). The remainder of the Missouri River species range at least as far upstream in the Mississippi River as Minnesota (Greene, 1935; Eddy and Surber, 1947), although dam construction in recent years has apparently impeded movement of the skipjack herring (*Alosa chrysochloris*) and the blue catfish (*Ictalurus furcatus*) so that these no longer occur so far north (Barnickol and Starrett, 1951: 305, 323; Eddy and Underhill, 1959). Among Missouri River fishes of South Dakota, listed below, those that range upstream in the Mississippi River into the Minnesota River drainage in that state or anywhere in the Red River basin are indicated by asterisks: **Ichthyomyzon unicuspis, Scaphirhynchus platorynchus, Polyodon spathula, Lepisosteus platostomus, *Lepisosteus osseus, *Alosa chrysochloris, Dorosoma cepedianum, *Hiodon alosoides, *Hybopsis storeriana, *Notropis atherinoides, *Notropis blennius, Hybognathus nuchalis, Cycleptus elongatus, *Ictiobus cyprinellus, Ictiobus bubalus, *Carpiodes carpio, *Ictalurus melas, *Ictalurus punctatus, Ictalurus furcatus, *Noturus flavus, Pylodictis olivaris, *Anguilla rostrata, *Lota lota, *Pomoxis annularis, *Stizostedion canadense, *Stizostedion vitreum,* and **Aplodinotus grunniens.*

The 23 remaining species that entered South Dakota postglacially by way of the Missouri River are not typical inhabitants of large, turbid streams; rather they usually occur in clearer, often vegetated backwaters, lakes, creeks, small rivers, and spring runs. For the following 16 species we have no South Dakota records in the Missouri River: *Notemigonus crysoleucas, Semotilus atromaculatus, Hybopsis biguttata, Rhinichthys cataractae, Phenacobius mirabilis, Notropis cornutus, N. topeka* (Fig. 6), *N. heterolepis, Hybognathus hankinsoni, Campostoma anomalum, Carpiodes cyprinus, Moxostoma macrolepidotum, Noturus gyrinus, Fundulus sciadicus, Culaea inconstans,* and *Etheostoma exile.* Our collections include South Dakota stations in the Missouri River or closely adjacent waters for *Notropis dorsalis, N. lutrensis, N. stramineus missuriensis, Pimephales promelas,*

Catostomus commersoni, Lepomis cyanellus, and *L. humilis.* These are interpreted as based on stragglers temporarily outside their usual haunts, but illustrate the method whereby non-river species utilize large streams in dispersal. It should be noted that the high turbidity characteristic of the Missouri River in warm months is not present throughout the winter. To what extent clear-water fishes move through the river then is not known. Many of the non-river species live only in the southern or southeastern part of the state (Table 9), a limitation that for some may result from the barrier to dispersal posed by the Missouri River.

UPPER MISSISSIPPI RIVER.—Fifty-five fishes have employed connecting waters of the Upper Mississippi, chiefly the Minnesota River and its tributaries, to gain access to South Dakota (Table 9). Fifty of these are known to live in the Minnesota River drainage of the state, and the others all occur elsewhere in that drainage. Most fishes moving northward utilized River Warren to enter the basin of Lake Agassiz, now the Red River drainage (p. 113). Of these, 18 have been collected in South Dakota. The chief faunal difference between the fishes of the Minnesota and Red rivers is one of reduction and depends on the time of their arrival during northward dispersal. Species that arrived too late to enter the Red basin include *Amia calva, Alosa chrysochloris, Campostoma anomalum, Hypentelium nigricans, Moxostoma erythrurum, Noturus flavus, Anguilla rostrata, Roccus chrysops, Micropterus dolomieui, Micropterus salmoides, Lepomis macrochirus, Lepomis humilis, Pomoxis annularis, Pomoxis nigromaculatus,* and *Percina phoxocephala.* To these South Dakota species may be added *Etheostoma zonale, E. caeruleum,* and *E. flabellare* from the upper Minnesota River basin of Minnesota (Underhill, 1957: 29). Recent introductions (Table 9), chiefly of warm-water species such as the centrarchids, have dulled the sharpness of the faunal barrier. Eight fishes occur in South Dakota only in the Minnesota and/or the Red River basins: *Amia calva, Notropis rubellus, Notropis heterodon, Hypentelium nigricans, Moxostoma erythrurum, Ictalurus nebulosus, Micropterus dolomieui,* and *Percina phoxocephala.*

AN UPPER MISSISSIPPI BASIN COMPONENT IN THE ICHTHYOFAUNA OF THE MISSOURI RIVER DRAINAGE.—A group of fishes living in the Little Sioux, Big Sioux, Vermillion, and James rivers, consists of forms characteristic of the upper Mississippi drainage that are absent or are much restricted in occurrence elsewhere in the Missouri basin. This distributional pattern could be achieved in one or more ways: (1) human introduction, (2) reinvasion of this area from the Missouri basin following the Wisconsin glaciation and in time by disappearance or notable reduction from the glacial refugia to the south and west, (3) by a major early postglacial drainage readjustment that brought Mississippi species into the Missouri drainage, for example via the Red and James rivers in North Dakota, or (4) by headwater stream connections between the upper Mississippi and upper Missouri drainages. It has been indicated above (p. 113) that the third alternative is not supported by geological evidence. Perhaps more than one source has been involved, but it is our contention that the fourth explanation is primarily responsible for bringing the following 23 fishes into that part of the Missouri drainage that includes eastern North and South Dakota, southwestern Minnesota, northwestern Iowa, and northeastern Nebraska (those forms not yet known from the South Dakota part of the area are indicated by asterisks): *Umbra limi, Esox lucius, Chrosomus eos* (in Big Sioux drainage), *Rhinichthys atratulus meleagris, Notropis hudsonius hudsonius,* **Notropis anogenus,* **Notropis heterodon, Notropis stramineus stramineus, Pimephales notatus,* **Moxostoma erythrurum, Fundulus diaphanus menona, Percopsis omiscomaycus, Roccus chrysops,* **Micropterus dolomieui, Micropterus salmoides, Lepomis gibbosus, Lepomis macrochirus, Ambloplites rupestris rupestris, Pomoxis nigromaculatus, Perca flavescens, Percina maculata, Percina caprodes semifasciata,* and *Etheostoma nigrum.*

That human introduction alone could have been responsible for this distributional phenomenon is incredible. The pattern is too broad, it involves inconsequential non-game as well as important game and food fishes, and, most important, early collectors found this same assortment of species prior to the extensive stocking programs of more recent time. For some species, however, early transfers of game fishes, especially centrarchids, may have confused distribution. In 1892 Meek wrote (p. 219): "Mr. B. F. Shaw, at one time fish commissioner of Iowa, did very effective work during his occupancy of that office in seining the fishes out of many of these [Mississippi River] bayous, where a great mortality occurs annually, and depositing them in the lakes and rivers . . ." Thus, even at that early date the possibility of an introduction, particularly of a food fish, is not excluded.

A logical explanation for a distribution such as that described is to postulate a Pleistocene range well beyond the glacial front (in this situation to the west or south) with postglacial redispersal into the glaciated area followed by withdrawal (or extinction) from the previous range. This explanation depends on the assumption that all of the previously habitable refugium has been deserted as unsuited for occupancy, a displacement which surely befell some animals after glacial recession. It is usual, however, to encounter at least isolated relict populations in favorable habitats following dispersal. The unacceptability of this postulate here depends (a) on the general distributional patterns of the included species, (b) on intraspecific geographic replacement in and out of the glaciated area, and (c) on the apparent availability of a postglacial avenue of dispersal that more plausibly satisfies the zoogeographic findings.

The 23 species listed are found in the Missouri basin wholly or largely in the contiguous drainage systems of the James, Vermillion, Big Sioux, and Little Sioux rivers, and minor intervening streams. A few have apparently spread somewhat to the south or west from this area, but such movement has been mainly downstream. For example, 20 forms occur in the Big Sioux drainage, 10 are in the James drainage, 19 are in the Little Sioux drainage, but only *Pimephales notatus* lives in the Elkhorn River of Nebraska. *Rhinichthys atratulus meleagris* ranges a short distance upstream from the James River in the Missouri drainage, but no other form is known to have been present originally in the Missouri drainage above the James. (*Esox lucius, Pimephales notatus, Fundulus diaphanus menona, Micropterus salmoides, Lepomis gibbosus, Lepomis macrochirus, Ambloplites rupestris, Pomoxis nigromaculatus,* and *Perca flavescens* have been established in the upper Missouri through introduction, usually in localized areas.) Although with minor exceptions these 23 species are much restricted in their occurrence in the Missouri basin, all occur in the upper Mississippi basin and 16 are in the Red River drainage. It seems appropriate, therefore, to inquire into the derivation of this faunal assemblage from the upper Mississipppi basin.

Intraspecific replacement of stocks provides perhaps the most graphic demonstration supporting a postglacial faunal invasion of part of the Missouri basin from the north and east. Two species supply evidence on this. The widespread sand shiner of the Great Plains, including most of the Missouri basin, is *Notropis stramineus missuriensis* (Table 5; p. 66). The typical subspecies, *stramineus,* enters the lower part of the Missouri basin and intergrades with *missuriensis* in northwestern Missouri and in the Big and Little Nemaha river system of southeastern Nebraska (Raymond E. Johnson, MS). Following glacial withdrawal from eastern South Dakota and adjacent regions in the basin, one would expect invasion from the south by *missuriensis*. But analysis indicates that the James, Vermillion, Big Sioux, and Little Sioux drainages are occupied by *N. s. stramineus,* as are the Minnesota and Red rivers. *N. s. missuriensis* inhabits southern and western tributaries to the Missouri and those on the east bank as far south as Emanuel Creek in Bon Homme County, South Dakota, less than 30 miles west of the James River. The geographic pattern points to a transfer of *N. s. stramineus* into the middle Missouri basin, from the Red, Minnesota, or Des Moines river basin.

Etheostoma nigrum is represented in the northcentral states by two forms that may be distinguished in squamation, in habitat, and in distribution (Greene, 1935; Hubbs and Greene, 1935; Lagler and Bailey, 1947). It is evident from examination of many lots that over certain considerable areas the stocks of this species are phenotypically intermediate between the subspecies *eulepis* and *nigrum*. One such area includes the Big Stone Lake basin, presumably the Minnesota River, and the Mississippi River from St. Paul to St. Louis. The typical subspecies occurs in the lower part of the Missouri basin. In South Dakota the species occurs in the James, Vermillion, and the Big Sioux, as well as in the Minnesota drainage and all samples are identifiable as intermediate between the subspecies (Table 8). Again the evidence points to derivation from the upper Mississippi system.

It has been indicated (pp. 113–14) that there is sound evidence of hydrographic interchange between the Minnesota and Big Sioux basins, and former intercommunication between the Des Moines and Little Sioux rivers probably occurred. In the Missouri basin the rivers from the James to the Little Sioux are close together so that limited interchange of their fishes is possible. We believe that the data presented point to faunal transfer from the Mississippi to the Missouri, but movement in the opposite direction is not precluded. Perhaps *Notropis topeka* entered the Des Moines basin in this way (Fig. 6). Interchange of fish stocks was certainly not restricted to the forms listed. Some species (e. g., *Semotilus atromaculatus, Notropis cornutus, Pimephales promelas, Catostomus commersoni*) probably re-entered the area from both pathways, and the details of adjustment of populations cannot readily be deciphered.

INTRODUCED SPECIES.—Six species are known, or thought, to have been introduced into South Dakota through direct human intervention. *Salmo trutta, Salmo gairdneri, Salvelinus fontinalis, Cyprinus carpio,* and *Carassius auratus* are assuredly exotic, and *Fundulus kansae* was likely introduced into the Cheyenne River system (Miller, 1955: 11–12), presumably from the Platte River drainage where it is native. Several species native to some South Dakota waters have had their ranges extended through introduction (Table 9); for some it is impossible now to delineate original ranges with precision.

LITERATURE CITED

ALLUM, MARVIN O., AND ERNEST J. HUGGHINS
    1959 Epizootics of fish lice, *Argulus biramosus*, in two lakes of eastern South Dakota. (Abstr.) Jour. Parasitol., 45 (4–Sec. 2) : 33.

BAILEY, REEVE M.
    1951 A check-list of the fishes of Iowa, with keys for identification, Pp. 185–237. *In* James R. Harlan and Everett B. Speaker, Iowa fish and fishing. Iowa Conserv. Comm., Des Moines.
    1954 Distribution of the American cyprinid fish *Hybognathus hankinsoni* with comments on its original description. Copeia, (4): 289–91.
    1956 A revised list of the fishes of Iowa, with keys for identificaton, Pp. 327–77. *In* James R. Harlan and Everett B. Speaker, Iowa fish and fishing. Iowa Conserv. Comm., Des Moines.
    1959a Distribution of the American cyprinid fish *Notropis anogenus*. Copeia, (2): 119–23.
    1959b Parasitic lampreys (*Ichthyomyzon*) from the Missouri River, Missouri and South Dakota. Copeia, (2): 162–63.

BAILEY, REEVE M., AND FRANK B. CROSS
    1954 River sturgeons of the American genus *Scaphirhynchus*: characters, distribution, and synonymy. Papers Mich. Acad. Sci., 39: 169–208.

BARNICKOL, PAUL G., AND WILLIAM C. STARRETT
    1951 Commercial and sport fishes of the Mississippi River between Caruthersville, Missouri, and Dubuque, Iowa. Bull. Illinois Nat. Hist. Survey, 25: 267–350.

BROWN, C. J. D.
    1951 The paddlefish in Fort Peck Reservoir, Montana. Copeia, (3): 252.
    1955 A record-size pallid sturgeon, *Scaphirhynchus album* from Montana. Copeia, (1): 55.

BURROUGHS, RAYMOND DARWIN
    1961 The natural history of the Lewis and Clark Expedition. Mich. State Univ. Press, East Lansing. xii + 340 pp.

CHURCHILL, E. P., JR.
    1927 Distribution of certain newly recorded fish of South Dakota. Bull. Ecol. Soc. Amer., 8: 6–7.

CHURCHILL, EDWARD P., AND WILLIAM H. OVER
    1933 Fishes of South Dakota. S. D. Dept. Game and Fish, Pierre. 87 pp.

CLEARY, ROBERT E.
    1956 The distribution of the fishes of Iowa, Pp. 267–324, Maps 1–98. *In* James R. Harlan and Everett B. Speaker, Iowa fish and fishing. Iowa Conserv. Comm., Des Moines.

COPE, E. D.
    1871 Report on the reptiles and fishes obtained by the naturalists of the expedition, Pp. 432–42. *In* F. V. Hayden, Second [fourth] annual report U. S. Geological Survey of Wyoming and portions of contiguous territories.
    1879 A contribution to the zoology of Montana. Amer. Nat., 13: 432–41.

Cox, Ulysses O.
- 1896 A report upon the fishes of southwestern Minnesota. Rept. U. S. Comm. Fish and Fish., 20 (1894): 605–16.
- 1897 A preliminary report on the fishes of Minnesota. Geol. and Nat. Hist. Survey Minn., Zool. Ser., 3: 1–93.

Cross, Frank B.
- 1953 Occurrence of the sturgeon chub, *Hybopsis gelida* (Girard) in Kansas. Trans. Kans. Acad. Sci., 56: 90–91.

Eddy, Samuel
- 1957 How to know the freshwater fishes. W. C. Brown Co., Dubuque, Iowa. 276 pp.

Eddy, Samuel, and Thaddeus Surber
- 1943 Northern fishes, with special reference to the Upper Mississippi Valley. Univ. Minn. Press, Minneapolis. xii + 252 pp.
- 1947 Northern fishes, with special reference to the Upper Mississipppi Valley. Rev. ed. Univ. Minn. Press, Minneapolis. xii + 276 pp.

Eddy, Samuel, and James C. Underhill
- 1959 Recent changes and corrections for the Minnesota fish fauna. Copeia, (4): 342–43.

Evermann, Barton W.
- 1893a Description of a new sucker, *Pantosteus jordani*, from the upper Missouri basin. Bull. U. S. Fish Comm., 12 (1892): 51–56.
- 1893b The ichthyologic features of the Black Hills region. Proc. Indiana Acad. Sci., (1892): 73–78.

Evermann, Barton W., and Ulysses O. Cox
- 1896 A report upon the fishes of the Missouri River basin. Rept. U. S. Comm. Fish and Fish., 20 (1894): 325–429.

Evermann, Barton W., and J. T. Scovell
- 1896 Notes on a collection of fishes from the Missouri River basin. Proc. Indiana Acad. Sci., (1895): 126–30.

Fenneman, Nevin M.
- 1938 Physiography of eastern United States. McGraw-Hill Book Co., New York. 714 pp.

Flint, Richard Foster
- 1955 Pleistocene geology of eastern South Dakota. U. S. Geol. Survey Prof. Paper, 262: vi + 173 pp.

Forbes, Stephen Alfred, and Robert Earl Richardson
- 1909 (and ed. 2, 1920) The fishes of Illinois. Nat. Hist. Survey Illinois, 3: cxxxi + 357 pp.

Fowler, Henry W.
- 1915 Notes on nematognathous fishes. Proc. Acad. Nat. Sci. Philadelphia, 67: 203–43.

Gerking, Shelby D.
- 1945 The distribution of the fishes of Indiana. Invest. Indiana Lakes and Streams, 3 (1): 1–137.

GIBBS, ROBERT H., JR.
  1957 Cyprinid fishes of the subgenus *Cyprinella* of *Notropis*. II. Distribution and variation of *Notropis spilopterus*, with the description of a new subspecies. Lloydia, 20: 186–211.

GILBERT, CARTER R.
  1961 Hybridization versus intergradation: An inquiry into the relationship of two cyprinid fishes. Copeia, (2): 181–92.

GIRARD, CHARLES
  1856 Researches upon the cyprinoid fishes inhabiting the fresh waters of the United States, west of the Mississippi Valley, from specimens in the museum of the Smithsonian Institution. Proc. Acad. Nat. Sci. Philadelphia, 8: 165–213.
  1858 Fishes, p. 1–400. *In* General report on the zoology of the several Pacific railroad routes. U.S. Pacific R. R. Survey 10 (4).

GREENE, C. WILLARD
  1935 The distribution of Wisconsin fishes. Wisc. Conserv. Comm., Madison. 235 pp.

HANKINSON, THOMAS L.
  1929 Fishes of North Dakota. Papers Mich. Acad. Sci., 10 (1928): 439–60.
  1932 Observations on the breeding behavior and habitats of fishes in southern Michigan. Papers Mich. Acad. Sci., 15 (1931): 411–25.

HARLAN, JAMES R., AND EVERETT B. SPEAKER
  1951 Iowa fish and fishing. Iowa Conserv. Comm., Des Moines, 184 pp.
  1956 Iowa fish and fishing. *Ibid.*, 377 pp.

HARRISON, HARRY M., AND EVERETT B. SPEAKER
  1954 An annotated list of the fishes in the streams tributary to the Missouri River in Iowa. Proc. Iowa Acad. Sci., 61: 511–23.

HILDEBRAND, SAMUEL F.
  1932 On a new cyprinoid from South Dakota. Jour. Wash. Acad. Sci., 22: 257–60.

HIPSCHMAN, DON
  1959 Department history, 1909–1959: 12–73. *In* Fiftieth annual report South Dakota Department of Game, Fish, and Parks, Pierre.

HUBBS, CARL L.
  1926 A check-list of the fishes of the Great Lakes and tributary waters, with nomenclatorial notes and analytical keys. Misc. Publ. Mus. Zool. Univ. Mich., 15: 1–77.
  1930 Materials for a revision of the catostomid fishes of eastern North America. *Ibid.*, 20: 1–47.
  1945 Corrected distributional records for Minnesota fishes. Copeia, (1): 13–22.
  1951 *Notropis amnis*, a new cyprinid fish of the Mississippi fauna, with two subspecies. Occ. Papers Mus. Zool. Univ. Mich., 530: 1–30.

HUBBS, CARL L., AND JOHN D. BLACK
  1947 Revision of *Ceratichthys*, a genus of American cyprinid fishes. Misc. Publ. Mus. Zool. Univ. Mich., 66: 1–56.

HUBBS, CARL L., AND KELSHAW BONHAM
  1951 New cyprinid fishes of the genus *Notropis* from Texas. Copeia, (1): 91–110.

HUBBS, CARL L., AND C. WILLARD GREENE
  1935 Two new subspecies of fishes from Wisconsin. Trans. Wisc. Acad. Sci., 29: 89–101.

HUBBS, CARL L., LAURA C. HUBBS, AND RAYMOND E. JOHNSON
  1943 Hybridization in nature between species of catostomid fishes. Contrib. Lab. Vert. Biol. Univ. Mich., 25: 1–76.

HUBBS, CARL L., AND KARL F. LAGLER
  1941 Guide to the fishes of the Great Lakes and tributary waters. Bull. Cranbrook Inst. Sci., 18: 1–100.
  1958 Fishes of the Great Lakes region. Rev. ed., *Ibid.*, 26: 1–213.

HUBBS, CARL L., AND MILTON B. TRAUTMAN
  1937 A revision of the lamprey genus *Ichthyomyzon*. Misc. Publ. Mus. Zool. Univ. Mich., 35: 1–109.

HUGGHINS, ERNEST J.
  1959 Parasites of fishes in South Dakota. S. D. State Coll. Agric. Exp. Sta. Bull., 484: 1–73.

JORDAN, DAVID STARR, AND BARTON WARREN EVERMANN
  1896 The fishes of North and Middle America. Bull. U. S. Natl. Mus., 47 (1): 1–1240.

JORDAN, DAVID STARR, BARTON WARREN EVERMANN, AND HOWARD WALTON CLARK
  1930 Check list of the fishes and fishlike vertebrates of North and Middle America north of the northern boundary of Venezuela and Colombia. Rept. U. S. Comm. Fish., 1928 (2): 1–670.

LAGLER, KARL F., AND REEVE M. BAILEY
  1947 The genetic fixity of differential characters in subspecies of the percid fish, *Boleosoma nigrum*. Copeia, (1): 50–59.

LANGLOIS, T. H.
  1929 Breeding habits of the northern dace. Ecology, 10: 161–63.

LARRABEE, AUSTIN P.
  1926 An ecological study of the fishes of the Lake Okoboji region. Univ. Iowa Studies Nat. Hist., 11 (12): 1–35.

LEMKE, RICHARD W.
  1960 Geology of the Souris River area North Dakota. U. S. Geol. Surv. Prof. Paper, 325:ix + 138 pp.

LEVERETT, FRANK
  1932 Quaternary geology of Minnesota and parts of adjacent states. U. S. Geol. Survey Prof. Paper, 161: v + 149 pp.

LINDSEY, C. C.
  1956 Distribution and taxonomy of fishes in the Mackenzie drainage of British Columbia. Jour. Fish. Res. Bd. Canada, 13: 759–89.

LUDLOW, WILLIAM
  1875 Report of a reconnaissance of the Black Hills of Dakota made in the summer of 1874. Engr. Dept. U. S. Army, Washington, D. C., 121 pp.

ROBINS, C. RICHARD, AND EDWARD C. RANEY
  1957 Distributional and nomenclatorial notes on the suckers of the genus *Moxostoma*. Copeia, (2): 154–55.

SHIELDS, JAMES T.
  1958a Experimental control of carp reproduction through water drawdowns in Fort Randall Reservoir, South Dakota. Trans. Amer. Fish. Soc., 87 (1957): 23–33.
  1958b Fish management problems of large impoundments on the Missouri River. Ibid. 356–62.

SIMON, JAMES R.
  1946 Wyoming fishes. Wyo. Game and Fish Dept., Bull., 4: 1–129.

SUCKLEY, GEORGE
  1874 On the North American species of salmon and trout. Rept. U. S. Comm. Fish and Fish., 2 (1872 and 1873): 91–160.

SUTTKUS, ROYAL D.
  1958 Status of the nominal cyprinid species *Moniana deliciosa* Girard and *Cyprinella texana* Girard. Copeia, (4): 307–18.

TAKAHASI, NISUKE
  1925 On the homology of the cranial muscles of the cypriniform fishes. Jour. Morphol. and Physiol., 40: 1–109.

TRAUTMAN, MILTON B.
  1956 *Carpiodes cyprinus hinei*, a new subspecies of carpsucker from the Ohio and Upper Mississippi river systems. Ohio Jour. Sci., 56: 33–40.
  1957 The fishes of Ohio, with illustrated keys. Ohio State Univ. Press, Columbus, xviii + 683 pp.

TRAUTMAN, MILTON B., AND ROBERT G. MARTIN
  1951 *Moxostoma aureolum pisolabrum*, a new subspecies of sucker from the Ozarkian streams of the Mississippi River system. Occ. Papers Mus. Zool. Univ. Mich., 534: 1–10.

UNDERHILL, JAMES C.
  1957 The distribution of Minnesota minnows and darters. Minn. Mus. Nat. Hist. Univ. Minn. Occ. Papers, 7: 1–45.
  1959 Fishes of the Vermillion River, South Dakota. Proc. S. D. Acad. Sci., 38: 96–102.

UPHAM, WARREN
  1896 The glacial Lake Agassiz. U. S. Geol. Survey Monograph, 25: 1–658.

WHITLEY, G. P.
  1950 New fish names. Proc. Royal Zool. Soc. New S. Wales, 1948–49: 44.

WOOLMAN, ALBERT J.
  1896 Report upon the ichthyological investigations in western Minnesota and eastern North Dakota. Rept. U. S. Comm. Fish and Fish., 19, 1893 (1895): 343–73.

Accepted for publication November 9, 1961

MARTIN, W. R.
    1949 The mechanics of environmental control of body form in fishes. Univ. Toronto Studies Biol. Ser., 58, Publ. Ontario Fish. Res. Lab., 70: 1–91.

MEEK, SETH EUGENE
    1892 A report upon the fishes of Iowa, based upon observations and collections made during 1889, 1890, and 1891. Bull. U. S. Fish Comm., 10 (1890): 217–48.
    1894 Notes on the fishes of western Iowa and eastern Nebraska. Bull. U. S. Fish Comm., 14: 133–38.

MILLER, ROBERT RUSH
    1955 An annotated list of the American cyprinodontid fishes of the genus *Fundulus*, with the description of *Fundulus persimilis* from Yucatan. Occ. Papers Mus. Zool. Univ. Mich., 568: 1–25.

MINCKLEY, W. L., AND FRANK B. CROSS
    1959 Distribution, habitat, and abundance of the Topeka shiner *Notropis topeka* (Gilbert) in Kansas. Amer. Midland Nat., 61: 210–17.
    1960 Taxonomic status of the shorthead redhorse, *Moxostoma aureolum* (LeSueur) from the Kansas River basin, Kansas. Trans. Kans. Acad. Sci., 63: 35–39.

MOORE, GEORGE A.
    1950 The cutaneous sense organs of barbeled minnows adapted to life in the muddy waters of the Great Plains region. Trans. Amer. Microscop. Soc., 69: 69–95.
    1957 Fishes, pp. 32–210. *In* W. F. Blair *et al.*, Vertebrates of the United States. McGraw-Hill Book Co., New York.

MOYLE, JOHN B., AND WILLIAM D. CLOTHIER
    1959 Effects of management and winter oxygen levels on the fish population of a prairie lake. Trans. Amer. Fish. Soc., 88: 178–85.

OLUND, LEONARD J., AND FRANK B. CROSS
    1961 Geographic variation in the North American cyprinid fish, *Hybopsis gracilis*. Univ. Kans. Publ. Mus. Nat. Hist., 13: 323–48.

OVER, WILLIAM H., AND EDWARD P. CHURCHILL
    1927 A preliminary report of a biological survey of the lakes of South Dakota. S. D. Geol. and Nat. Hist. Survey, Circ. 29 (6): 1–17.

PERSONIUS, ROBERT GILES, AND SAMUEL EDDY
    1955 Fishes of the Little Missouri River. Copeia, (1): 41–43.

RANEY, EDWARD C.
    1940 The breeding behavior of the common shiner, *Notropis cornutus* (Mitchill). Zoologica, 25: 1–14.
    1949 Nests under the water. Canadian Nature, 11: 71–78.

REIGHARD, JACOB
    1910 Methods of studying the habits of fishes with an account of the breeding habits of the horned dace. Bull. U. S. Bur. Fish., 28 (1908), pt. 2: 1111–36.
    1943 The breeding habits of the river chub, *Nocomis micropogon* (Cope). Papers Mich. Acad. Sci., 28: 397–423.

PLATE I

Pharyngeal apparatus in *Hybognathus placitus* and *H. n. nuchalis*

All figures in ventral view, anterior at top. *bo*, posterior process of basioccipital; *g*, gill; *pa*, pharyngeal arches; *p pad*, pharyngeal pad; *p pro*, pharyngeal process of basioccipital; *pt*, pharyngeal tooth; *r*, retractor arcus branchialis dorsalis inferior; *tr*, transversus ventralis.

A. *H. placitus*. Dissection to expose pharyngeal structures in situ. UMMZ 94799, 77 mm. s. l., Little Missouri River, Marmarth, North Dakota.

B. *H. placitus*. Basioccipital; the pharyngeal pad has been removed. UMMZ 94799, 82 mm. s. l.

C. *H. n. nuchalis*. Dissection to expose pharyngeal structures in situ. UMMZ 94800, 83 mm. s. l., Little Missouri River, below Marmarth, North Dakota.

D. *H. n. nuchalis*. Basioccipital; the pharyngeal pad has been removed. UMMZ 94800, 87 mm. s. l.

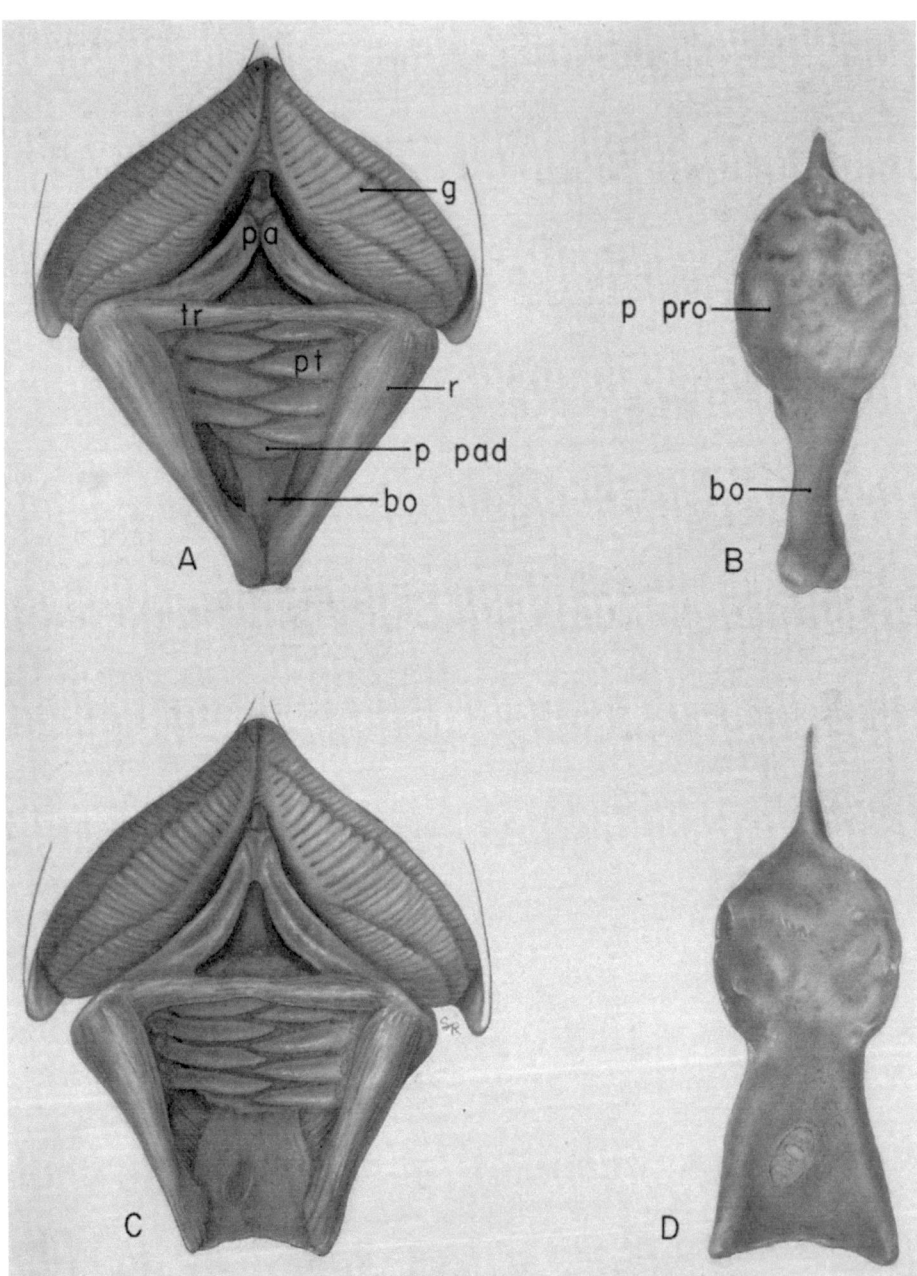